FROM
WATTLE & DAUB
TO
CONCRETE & STEEL

66.00

# FROM WATTLE & DAUB TO CONCRETE & STEEL

## THE ENGINEERING HERITAGE OF AUSTRALIA'S BUILDINGS

Henry J. Cowan AO

Illustrated by
Peter R. Smith

MELBOURNE UNIVERSITY PRESS

Melbourne University Press
PO Box 278 Carlton South, Victoria 3053, Australia
info@mup.unimelb.edu.au
www.mup.com.au

First published 1998



Designed and typeset by Lauren Statham
Typeset in 10.5 point Plantin Light
Printed in Australia by Australian Print Group

National Library of Australia Cataloguing-in-Publication entry

Cowan, Henry J. (Henry Jacob), 1919– .

From wattle and daub to concrete and steel: the engineering heritage of Australia's buildings.

Bibliography.
Includes index.
ISBN 0 522 84730 7.

1. Building—Australia—History. 2. Buildings—Australia—History. 1. Title.

690.0994

To Judith and David

# FOREWORD

SINCE THE EARLY 1970s Professor H. J. Cowan has exercised significant influence in Australia in developing and increasing our awareness of the technical aspects of buildings. His latest book *From Wattle and Daub to Concrete and Steel* continues and adds to the tradition established in his earlier work.

Few people in Australia were interested in the preservation of old buildings before the end of the World War II, and the heritage movement gathered momentum only in the 1960s when some fine old buildings on valuable land were demolished to make room for taller buildings. The new interest in preservation and restoration led to the realisation that Australian engineering and architectural design and construction, while mostly based on British and American precedent, had a history of its own. Professor Cowan's book attempts to trace the principal steps in its development from the primitive construction of the late eighteenth century to the scientifically-based methods of the post World War II era.

The history is an important subject in its own right, because even the most ambitious and financially well-supported preservation programme cannot put more than a small part of the building on view. We cannot see the timber roofs of most buildings, although the Mint Museum enterprisingly shows its own on a video set. We cannot see the iron and steel of a building's structure, because it is covered by other materials. The original hydraulic lifts have all gone; the pipes that provided the power are still under the city pavements, but there is nothing visible in Melbourne, except some cover plates, and little in Sydney.

A knowledge of history is also useful for future design. Innovations are generally intended to be improvements on existing techniques, but the results do not always come up to expectation. The weathering properties of traditional brick and stone buildings provided a useful lesson in the 1970s, when the lack of durability of concrete and metal facades became evident; and complaints about the thermal environment of modern buildings are causing us to re-examine the concepts in use at the turn of the century.

The history of building construction therefore has a practical use, but it is also an interesting story. It is only briefly considered in books on the history of Australian engineering and of architecture, and Australia is not mentioned at all in any of the existing books on the history of building construction worldwide. The following chapters provide a fascinating introduction.

I am sure this book will fill a knowledge gap identified by the author, and provide very enjoyable reading for many, particularly practitioners in the various facets of the Australian building industry.

Alex McLachlan AM
BE, MBdgSc, HonFIE(Aust), FIMC
Past President (1988), The Institution of Engineers, Australia

# CONTENTS

# PREFACE

THIS VOLUME TRACES the development of the present sophisticated technology used for major Australian building projects from its primitive beginnings in the eighteenth century. Very few of the people who came with the First Fleet had any skill in building crafts, and they inevitably lacked a knowledge of the properties of Australian species of timber, many so surprisingly different from those of Europe and America.

These problems caused much hardship in the early years, but they were overcome remarkably quickly by learning from experience and by making use of migrant talent. Elizabeth Bay House and the Darlinghurst Court House, two buildings that would at that time have been noteworthy in any British town of comparable size, were built a mere half-century after the First Settlement.

In spite of the long time it took for information to reach Australia, major innovations were frequently adapted to local use within a few years of their development overseas. Prominent examples are corrugated iron, gas light, piped sewerage, hydraulic lifts, and reinforced concrete.

My review is confined to the main trends in Australian building construction, and I have omitted reference to a number of important buildings for that reason. The Sydney Opera House, Australia's best-known building, has remained without imitators in this country. The same applies to the small number of concrete shell structures built in the 1950s, and to the shells Dante Bini designed in the 1970s.

The recently built long-span steel roof structures, of which the Swimming Pool of the Melbourne Olympics of 1956 was an isolated early example, are not yet a part of history. Several important venues for the Sydney Olympics in the year 2000 are still to be built at the time of writing.

I hope this brief account provides a useful introduction to a potentially much larger field of study. I have dealt only selectively with the surviving buildings, and not at all with their restoration, because books have already been published on these topics. The emphasis is on the changes that have occurred in architectural and engineering design and construction, and on their historical background.

The Glossary explains the abbreviations and the technical terms used. I am indebted to Dr Peter Smith for the original drawings in this book, and to him and to the following for many helpful comments: Mr Frank Brady AM, Mr John Breen, Mr Michael Clarke, Professor Tom Heath, Mr Trevor Howells, Mr Peter Phillips, Mr David Rowe, Associate Professor Ross Thorne, and Professor Alex Wargon. My particular appreciation is due to my wife Renate, who read the entire script to correct mistakes and locate passages that might baffle a non-professional reader.

H.J.C.
Sydney, June 1996

# 1

# THE PRIMITIVE ERA

## Building resources and skills of the First Fleet people

The transportation of British convicts to North America had begun in 1717, partly to remove the convicts from Britain, and partly to provide labour for the colonies. Those sentenced to transportation were sold to shipping contractors, who in turn sold them to plantation owners. The plantation owners acquired the right to the convicts' labour for the term of their sentence (Martin 1978, p. 64), but they also had to provide food and housing.

American opposition to the transportation of convicts grew more intense about two years before the Declaration of Independence on 4 July 1776, and it resulted in its suspension on 1 April 1777. Thirty thousand convicts had been shipped to America in sixty years (Martin 1978, p. 64).

In 1779, while the American War of Independence was still in progress, the House of Commons passed a *Transportation Act*, and set up a committee to consider whether one thousand convicts a year could be transported overseas.

Joseph Banks, who gave evidence in the same year, was an amateur scientist who had accompanied Lieutenant James Cook on his expedition to observe the transit of Venus in Tahiti in 1769. On the way back Cook made the first voyage along the east coast of Australia, and named the country New South Wales. He made his first landing on 29 April 1770, in a bay that he called Stingray Bay; he later renamed it Botany Bay because of the great number of new plants that Banks had found there during the seven days of their stay.

Banks gave a favourable account: he observed that there was plenty of fish; while most of the soil was barren, sufficient rich soil could support a very large number of people; there was an abundance of timber and fuel; the grass was luxuriant, and he did not doubt that oxen and sheep, if carried there, would thrive and increase in number; and the climate was moderate (Martin 1978, p. 66). What he did not observe was that, while he found an abundance of plants he had never seen before, the trees commonly used for house construction in England and in America were missing at Botany Bay.

No action was taken at the time. In 1784 the House of Commons passed another *Transportation Act*, and the following year it set up a committee under the chairmanship of Lord Beauchamp to select the best site. Banks again gave evidence in favour of Botany Bay. He had become President of the Royal Society in 1778, an office to which he was re-elected repeatedly until his death in 1820, and he had the confidence of King George III, whom he frequently advised on gardening and on scientific matters.

However, other sites were considered. The Gambia River in tropical Africa was abandoned after Edmund Burke said that, as all the convicts would die a horrible death, it would be cheaper and kinder to hang them in England (Martin 1978, p. 155). The naval sloop *Nautilus* was sent to investigate Das Voltas Bay, a British outpost between Portuguese Angola and the Dutch Cape of Good Hope, but its officers reported that it was too arid for a settlement. No ship was sent to investigate Botany Bay because the round trip would have taken at least eighteen months, and the committee had full confidence in Banks's judgement. New Zealand was also considered, but Banks thought that the New South Wales 'Indians' posed a lesser security risk than the warlike Maoris.

On 18 August 1786 the Home Secretary, Viscount Sydney, who was responsible for both the colonies and the convicts, addressed a formal letter to the Lords Commissioners of the Treasury (*HRNSW* 1892, vol. 1, part 2, p. 14). He pointed out that, following the adverse report by the captain of the *Nautilus* on the southern coast of Africa, 'His Majesty thought it advisable to fix upon Botany Bay'.

Lord Sydney then set out in detail the vessels and men that would be needed 'for the conveyance of 750 convicts to Botany Bay', and the materials and equipment required, together with their estimated cost. The memorandum lists in detail the tools that should be carried:

> for each convict, 1 spade, 1 shovel, 3 hoes, 1 felling axe, 1 hatchet, and 1 knife;
> for each four men, 1 hand saw and 1 hammer;
> and in addition 40 cross-cut saws, 40 frame saws, 100 adzes, 100 broad axes, 140 augers, 140 drawing knives, 300 chisels and gouges, 100 planes, 10 iron forges, 10 anvils and hammers, 30 grindstones, 40 wheelbarrows, 50 pickaxes, 12 ploughs, and 40 iron hand mills.

It seems that the contractors decided that anything would do for convicts, and that the stores were not properly checked before departure, because Governor Arthur Phillip wrote to Home Under-Secretary Evan Nepean (*HRNSW* 1892, vol. 1, part 2, p. 154) on 9 July 1788, less than six months after landing, that 'most of the axes, spades, and shovels are the worst I have ever seen'. He enclosed a long list of tools he urgently needed, and he sent similar requests repeatedly during his four-year term (*HRNSW*, 1892, vol. 1, part 2, p. 154).

Lord Sydney's memorandum also lists the building materials to be sent: 10 barrels of nails of different sizes, 2000 spikes, 200 pairs of hinges, 100 locks, 10 tons of flat and square bar iron, and 1000 squares of crown glass (which was subsequently used in the First Government House, see p. 22).

Items not listed in the memorandum, but known to have been carried by the First Fleet, are a prefabricated government house (see p. 19), a few bags of lime, wooden moulds for making bricks, and some bricks that probably came as ballast. Various figures have been given for these; the most recent (Bridges 1995, p. 27) is 10 000.

It was assumed that adequate supplies of suitable timber, which would constitute the material for the first houses, would be found locally, as they had been in all the newly settled American colonies.

Two months later the Home Office asked for 500 tents for the convicts, and 160 for the marines (*HRNSW* 1892, vol. 1, part 2, p. 31).

Governor Phillip's commission, dated 12 October 1786, (*HRNSW* 1892, vol. I, part 2, p. 24) included 'the building of those castles, cities, boroughs, and fortifications' he thought necessary; but the First Fleet included only a few people with building experience.

Don Chapman (1981) compiled an almost complete list of the approximately 1300 people, convicts and escorts, who sailed with the First Fleet, and gave mini-biographies for each of them. There were twelve convict carpenters, but it seems that most were neither well-qualified nor hardworking. Phillip almost immediately hired all the sixteen ship's carpenters to build the first huts.

Eight years later Governor John Hunter was still hiring ships' carpenters, and was apparently reprimanded for using them instead of unpaid convict labour, because he wrote to the Home Secretary, the Duke of Portland, on 20 August 1796 that

> the necessities I find myself under, in consequence of the scarcity of artificers, and the want of public buildings indispensably requisite for the preservation of crops and other essential purposes, of hiring all soldiers and other free people that I could find; by this we shall, though unavoidably, incur some expense. (*HRA* 1914, vol. 1, p.59)

The position was more favourable for brickmaking and laying. There were two convict brickmakers, both competent, (see p. 12), and two convict bricklayers; one of the bricklayers, James Bloodsworth, is mentioned in most histories of Australia's First Settlement (see p. 22). This is at least one reason why bricks became such an important material in the construction of the early buildings. There was also one mason among the marines, and one convict plasterer.

There was an equal shortage of adequate supervisory expertise. It was not unusual for the army and the navy to undertake construction by direct labour, and several of the officers must therefore have had experience of supervising building operations. Lieutenant William Dawes, the artillery officer, was a skilled draughtsman (Figure 1).

In addition Phillip hired two men with building experience with whom he had worked before: Augustus Alt as Surveyor-General, and Henry Brewer, a fifty-year-old Clerk of Works, who was given the nominal rank of Midshipman.

When convicts had been transported to America, the local planters had provided the infrastructure, and required only unskilled labour. When colonies had previously been founded by free migrants in places whose resources were unknown, the organisers endeavoured to select those who really wanted to go, and a fair proportion of people with agricultural and building skills.

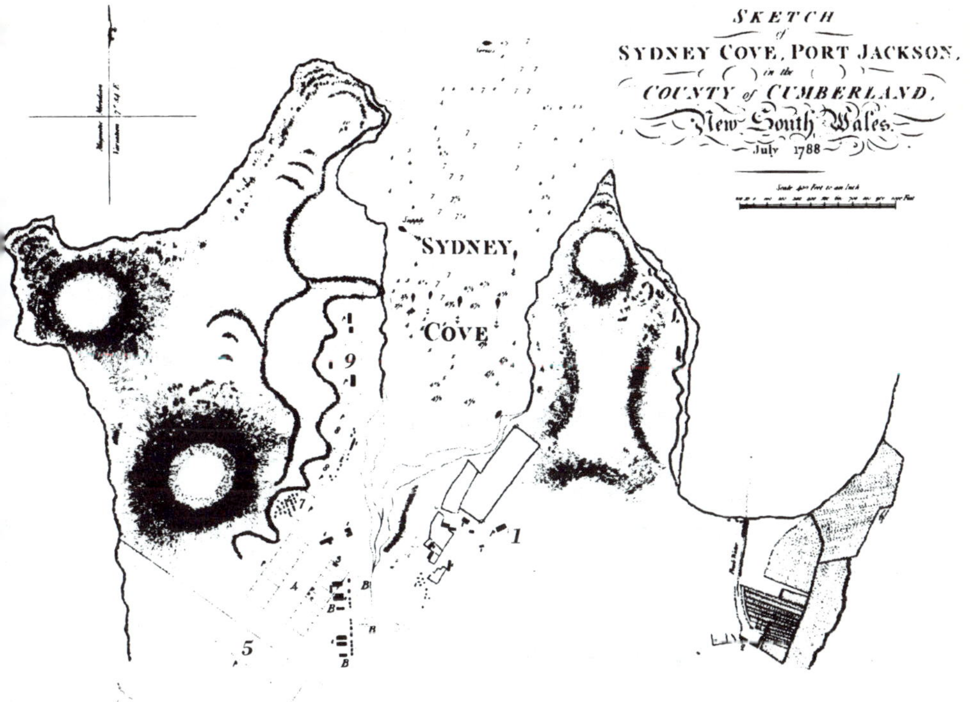


**Figure 1** Map of Sydney Cove drawn by Lieutenant William Dawes of the Marines in July 1788, six months after the First Settlement. The numbers denote:

1 A small House building for the Governeur.

5 Ground intended for the Governeur's House.

9 The Hospital.

The lower part of the map, which is not reproduced, also shows a town plan by Surveyor-General Alt with a 200 foot (60 m) wide street, which was not implemented.

(© Reproduced from a print published in 1983 by the Land Information Centre, Panorama Avenue, Bathurst 2795, from an original print in the Mitchell Library)

The problem was appreciated before the First Fleet sailed. *The Daily Universal Register* (later called the *Times*) editorialised already in 1786 that the lack of skill among the convicts threatened the new settlement with disaster. It proposed that bricklayers, carpenters and smiths should be sent out as free settlers and 'given some indulgence' (Martin 1978, p. 173).

The Home Office decided against sending free settlers with the First Fleet; Chapman's list (1986, p. 166) includes only one, and that man was found to lack the skills he had professed. Several of the officers who sailed with the First Fleet kept diaries, and some published books on their return. The most interesting of these are the two books by Watkin Tench, who sailed with the First Fleet as a captain of marines, and returned after four years when the marines were withdrawn. He later became Commandant of the Marine Corps with the rank of Lieutenant-General. In his second book, published in 1793, he wrote:

> Of the policy of settling, with convicts only, a country at once so remote and so extensive, I shall offer no remarks. Whenever I have had this question agitated, since my return to England, the cry of 'what can we do with them! where else can they be sent!' has always silenced me. (Tench 1961, p. 262)

This, then, was the main reason why building construction in Australia was more primitive than it had previously been in America, or subsequently in New Zealand. It is only fair to point out that food production proved even more difficult than building construction. During the first few years people were better housed than fed.

## The search for suitable timber

The lack of skill and enthusiasm among the convicts was one major problem. The lack of familiar species of timber was another.

Phillip decided almost immediately that Botany Bay was unsuitable for a settlement, and explored Port Jackson, which Cook had named after the then-Secretary of the Admiralty but had not entered. Now called Sydney Harbour, it was deep enough for ships to anchor close to the shore, and it had several sheltered bays. Phillip chose one (which he named Sydney Cove) that had a small stream, subsequently called the Tank Stream (now an underground stormwater sewer), and a good stand of tall straight trees.

The tall straight trees proved a disappointment. They were Sydney red gum (*Angophora costata*), a species in which the heart of the tree frequently rots while it is still young, leaving only a shell of sound wood. Sawpits were dug to cut up those that appeared usable, but as there was no time to season the timber the boards warped. Governor Phillip wrote in a memo to Lord Sydney:

> The timber is well described in Captain Cook's Voyage, but unfortunately it has one very bad quality, which puts us to great inconvenience; I mean the large gum-tree, which splits and warps in such a manner when used green, and to which necessity obliges us, that a store house boarded

up with this wood is rendered useless. (*HRNSW* 1892, vol. 1, part 2, p. 127)

Three years later Lieutenant Bradley listed twelve useful timbers in his Journal; the first of these is the red gum: 'Fit for little but the fire, but very good for that' (Bradley 1969, p. 232A).

Newton Fowell, a midshipman, described in his diary early slab huts that proved slightly more successful:

> The Convicts were constantly employed clearing Ground, building Store houses for the Reception of Provisions &c which were built by putting trees about 2 feet [0.6 m] in the ground so as to touch each other & thatched with rushes. (Fowell 1988, p. 75)

The rushes grew abundantly in Rushcutters Bay.

Not far from Sydney Cove were numerous moist gullies in which cabbage-palm trees grew. These were easy to fell, transport, and cut, but the timber was not strong enough to be used on its own. A number of buildings were constructed with timber frames, first from sound red-gum trees, later from she-oak, sheathed with cabbage palm, and covered with clay plaster for insulation and waterproofing (Figure 2); no lime deposits had been found. This was a temporary solution as within a few

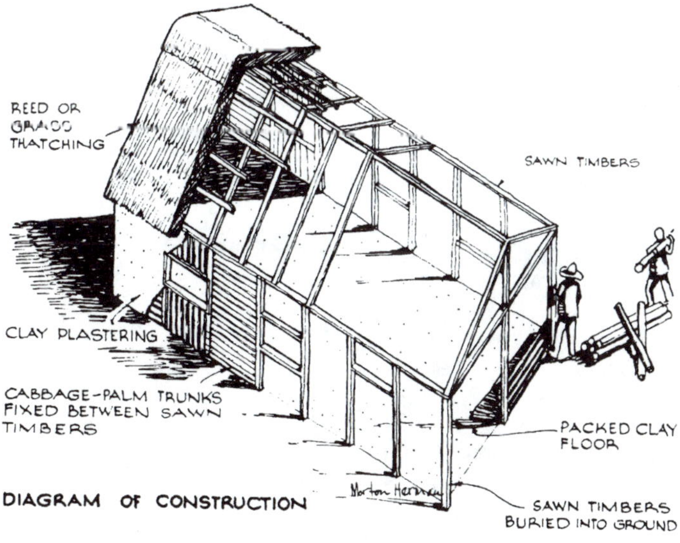


**Figure 2** Construction of the first huts built in Sydney, based on the book, *An Account of the English Colony of New South Wales*, published in London in 1798 by David Collins, the Judge Advocate of the First Settlement, and on a painting by Thomas Watling, made in 1793.

(Drawing by Morton Herman, reproduced from Herman [1970, p. 5], by permission of Harper Collins)

months all the cabbage palms near Sydney Cove and the neighbouring bays had been cut down, and the buildings did not last long because the cabbage-palm wood lacked durability.

Wattle-and-daub construction was more successful. The panels between the wooden posts were filled with mats woven, or 'wattled', from twigs, and then plastered with clay (Figure 3). Since the most suitable twigs near Sydney Cove came from the acacia, this is still known as a wattle tree. Wattle twigs were also used for filling the holes left for windows, since glass was reserved for the Governor's and a few special houses.

Wattle-and-daub construction, in spite of its lack of durability, remained in use when better construction methods became available, because it was cheap. It was used for some quite large buildings, including the first church, St Philip's, built in 1793, which stood for five years near the junction of Hunter and Bligh Streets (not the site of the present St Philip's Church), before being destroyed by fire, suspected to be arson. The site is now marked by a monument.

As far as timber was concerned, the First Fleet had been unlucky in the choice of Sydney Cove; but the Sydney red gum was not, as Phillip

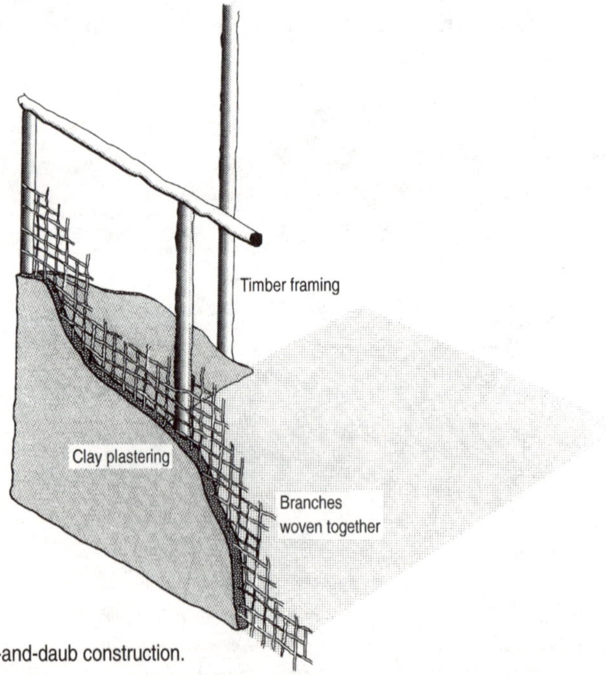


**Figure 3** Wattle-and-daub construction.
(Drawing by Peter R. Smith)

had at first assumed, a typical Australian tree. Several species were soon discovered that were suitable for building, although most had a tendency to warp on drying. The she-oak (*Casuarina fraseriana*), found during February 1788, split easily and could be used for roof shingles, which made roofs somewhat more waterproof and fire resistant than those of local thatch (see p. 16).

By July 1788 the peppermint gum (*Eucalyptus piperita*) was found, and shortly thereafter Sydney blue gum (*Eucalyptus saligna*), an excellent timber that was soon compared to English oak. Blackbutt (*Eucalyptus pilularis*), grey box (*Eucalyptus boisistoana*) and spotted gum (*Eucalyptus maculata*) came from Iron Cove (Cox & Freeland 1969, p. 26).

During exploration of the Hawkesbury region the red cedar (*Cedrela toona*) was found. It was an attractive-looking and easily-worked timber, which was used extensively for house-framing and joinery, as well as for furniture; it was even cut for fence-posts because it was so easy to work and the supply was initially plentiful. There were large stands in the coastal valleys, but it had become so popular that by 1870 there was very little left, in spite of an early attempt by the Governor at environmental control (Cox & Freeland 1969, p. 27).

By 1795, therefore, the problem of finding suitable timber supplies for building houses had largely been solved.

Eventually some of the Australian timber that had been found useless for building proved eminently suitable for street paving, and it provided a useful export as wood paving to England and as railway sleepers to India.

The problem of cutting the timber remained. Most of the useful species of timber so far discovered were so dense that they sank in water, and they were consequently hard to cut and hard to nail. If they were cut while green, they tended to warp; if they were seasoned, some became even harder.

The common method of sawing the timber during the first years was to dig a pit. Using a saw about 2 metres in length, one sawyer stood across the log and pulled the saw upwards with a horizontal handle, the other stood below the log in the pit and pulled downward on a vertical tiller. The sawdust fell on him, and he had a very unpleasant job working inside the pit, particularly in hot weather. However, pit-sawing continued to be used in remote areas of Australia until well into the nineteenth century.

The method was already obsolescent when the First Fleet sailed. The operation could be improved by mechanisms that converted the reciprocating into a circular movement, and a circular saw had been patented as early as 1777 (Elliott 1992, p. 9).

In a few suitable locations water or wind power was employed in the eighteenth century, and in the early nineteenth century steam power came into use. Peter Bridges (1995, p. 90) notes that a free settler brought a steam engine with him in 1813, which had a fitting that enabled it to be used to power a reciprocating saw, but it seems it was used only for a flour mill. Philip Cox and John Freeland (1969, p. 37) quote the *Sydney Morning Herald* of 31 October 1838, which reports the operation of the first steam-powered reciprocating saw in Australia earlier that month, but there is some reason to doubt that this was, in fact, the first use of a steam-powered saw in Australia.

William Charles Wentworth, the son of the surgeon D'Arcy Wentworth (see p. 27), wrote in 1819, while studying law in London, *A statistical, historical and political description of the Colony of New South Wales and its dependent settlements in Van Diemen's Land, with particular enumeration of the advantages which the Colonies can offer for emigration, and their superiority in many respects over those possessed by the United States of America* (*Australian Dictionary of Biography*, Vol. 2 1967). He was subsequently asked for advice by some intending emigrants, and suggested to several of them that, if they could afford it, they should include a mechanical saw in their equipment.

The first steam-powered circular saw started operating in Australia in the late 1830s or early 1840s. The use of power saws overcame the main problem posed by the hardness of the best Australian timbers, and by speeding up the conversion process they greatly reduced the cost of cut timber.

## Some other early settlements and the lime problem

Cook had observed the tall straight pine trees on Norfolk Island, and he had also seen flax growing there. In the era of sailing ships, timber tall and straight enough for masts and canvas for sails were constantly needed for replacement; they were as important to a sailing-ship navy as coal depots became when steam ships were introduced. The instruction of King George III to Governor Phillip required him *inter alia* 'to occupy Norfolk Island to prevent it being occupied by any other European power' (*HRNSW* 1892, vol. 1, Part 2, p. 85), and to secure these resources for the Royal Navy. He therefore sent a small garrison under the command of Lieutenant Philip King on 15 February 1788, a mere three weeks after arrival in Sydney Cove.

By that time the problems with the use of red gum for buildings had already become evident, and Governor Phillip instructed King to send a supply of pinewood to Sydney as soon as possible, in addition to finding some suitable for ships' masts. A number of Norfolk pines were felled, but found disappointing. They were tall and straight, but many had rotten cores, and the wood was too brittle for ships' masts.

King judged, however, that it would make good timber for building, and sent a consignment in August; Bridges (1995, p. 15) considered it probable that it was used in the First Government House (see p. 11). He also sent a small supply of limestone, of which there were numerous outcrops on the island. However, the Norfolk Island settlement lacked a port. In 1790 this resulted in the loss of the *Sirius*, which had been Phillip's flagship. In consequence Norfolk Island did not become a regular supplier of building materials.

In the meantime Sydney Cove was dependent for lime on the seashells that the women convicts collected, and later also on the Aboriginal middens found in various bays.

In 1791 the Hunter River, which provided a good port, was discovered, and a settlement was established in 1801. One of the tasks of the convicts was to collect shells to be sent to Sydney by boat.

In 1806 limestone was found near Port Dalrymple on the River Tamar, the second Tasmanian settlement, and in 1836 near Sorrento in the new settlement on Port Phillip Bay (Lewis 1988, p. 3). Some limestone also came from England as ballast. In 1813 the Blue Mountains were crossed, and in 1815 limestone was found in the interior; but even after a road had been built it was cheaper to bring lime by boat.

The shells gathered on the Hunter River and the Aboriginal middens found near it became a major source of lime supply (Gemmel 1986, p. 5) as the shell harvest in Sydney Harbour declined. However, Sydney's lime problem was solved only when the Southern Railway was built in the 1860s (Carroll 1988, p. 57), which gave economical access to the limestone of the interior. By 1880 rock lime had largely replaced shell lime.

## Brick and stone

The shortage of lime placed severe limitations on construction in stone and brick. In the colony's early days most mortar was made from clay or loam. This naturally reduced the strength of masonry and brickwork. Some mortar was given an admixture of hair to provide reinforcement. Loam with an addition of lime was used occasionally. Pure sand-lime mortar was utilised in the early days only in part of the First Government House; but its use became normal practice after adequate supplies of lime became available (see p. 11).

Stone was used immediately (see p. 21) for foundations; but only a few stone buildings were erected, in spite of the availability of sandstone, until lime mortar became more freely available (see pp. 24 and 51).

The Sydney sandstone is a freestone that can be cut accurately and relatively easily, and it would thus have been possible to use dry construction, which required perfectly fitting joints, since there was no mortar to fill the gaps. Greek stone masonry was generally laid dry, Roman masonry was sometimes laid dry, particularly for bridges where the use of wet mortar was inconvenient. The Inca masonry of Peru was laid dry, with joints so perfect that it is difficult to push a razor blade into them. However, dry construction would almost certainly have been unfamiliar to the mason and to the officers who came with the First Fleet.

Brick structures were used instead. The First Fleet brought two convict brickmakers, Samuel Wheeler and John King, and both seem to have been competent. Clay suitable for brickmaking was found quite soon about 1½ kilometres from Sydney Cove, near the junction of the present Liverpool and Pitt Streets, still called Brickfield Hill.

The brickmaking process was presumably similar to that used in England in the eighteenth century. A few wooden moulds had been brought by the First Fleet, but most of the tools had to be improvised. The sandy clay was ground with pestles made from tree trunks; water was then added, and the pug was worked by convicts with their bare feet. In the early days the pug was allowed to stand for only a few days, instead of the customary several weeks. The bricks and tiles were moulded by hand on a stock, a flat board 1 inch (25 mm) thick, and the size of the largest face of the brick. The excess was removed, and the surface of the wooden mould levelled with a strike board.

The mould was then removed, and the bricks were stacked with wood as a fuel, and covered with burnt bricks from an earlier firing, or in the early days with unburnt bricks. Because of the limited amount of air, the wood burnt slowly. The kiln was opened after about a week. This method did not alter significantly until the entire operation was mechanised (see p. 41), although the quality improved; but from 1840 the bricks were pressed into the moulds, which continued, however, to be handfilled.

The early colonial bricks were generally underburnt, because the highest temperature that could be attained was too low. They were consequently less strong and more porous than modern bricks; but they generally had an attractive light brownish-red colour.

During Governor Lachlan Macquarie's reign the British standard brick size was adopted, in the late eighteenth century 8¾ × 4 × 2½ inches (222 × 102 × 64 mm), but before that colonial bricks tended to be slightly larger (Pearson 1988, p. 5).

Captain Watkin Tench of the Marines wrote in November 1790:

> Wheeler, one of the master brick-makers, with two tile stools and one brick stool, was tasked to make and burn ready for use 30,000 tiles and bricks per month; he had 21 hands to assist him, who performed everything: cut wood, dug clay, &c. This continued, during the days of distress [the food shortage] excepted, when they did what they could, until June last. From June, with one brick and two tile stools he has been tasked to make 40,000 bricks and tiles monthly, as many of each sort as may be, having 22 men and 2 boys to assist him, on the same terms of procuring material as before. They fetch the clay of which tiles are made, two hundred yards; that for bricks is close at hand.
>
> He says that the bricks are such as in England would be called moderately good; and he judges that they would have fetched about 24 shillings per thousand at Kingston-upon-Thames, where he resided in the year 1784; their greatest fault is being too brittle. The tiles he thinks not so good as those made about London; the stuff has a rotten quality, and besides wants the advantage of being ground, in lieu of which they tread it.
>
> King, another master brick-maker, last year with the assistance of 16 men and 2 boys made 11,000 bricks weekly, with two stools. During the short allowance [of food] did what he could: resumed his old task when put again on full allowance; and had his number of assistants augmented to 20 men and 2 boys, on account of the increased distance of carrying wood for the kilns. He worked at Hammersmith for Mr. Scot, of that place. He thinks the bricks were as good as those made near London; and he says in the year 1784 they would have sold for a guinea [21 shillings] per thousand. (Tench 1961, p. 192)

In addition to these competent brickmakers there were at least two competent bricklayers, James Bloodsworth and Edward Parkins. Brick therefore became an important building material in the early days of the colony.

The term 'Colonial bond' is still used today for brickwork with one course of headers for each three courses of stretchers. However, the English bond (alternate courses of headers and stretchers) and Flemish bond (alternate headers and stretchers in each course) were also used. In the early buildings walls were generally thicker than the 9 inches (230 mm) that became customary later, because of the outward push by 'colonial' timber roofs (see p. 15). Walls were generally plastered on the outside, with clay in the early days, later with lime mortar, because the bricks were often porous and their surfaces uneven.

## Colonial roof construction

The annual rainfall at Sydney Cove is fairly high and the intensity of rainfall very high by comparison with that of London. Consequently few of the early Australian buildings had flat roofs, and hipped roofs providing sloping roof surfaces on all four sides were generally preferred to gabled roofs (Figure 4). Since the roof covering was not as waterproof as it became later, roof slopes in the early days were steep (Figure 8).

The roof structure in most of the early houses was primitive. It consisted simply of two heavy rafters resting against each other at the top, and then resting on top of the walls at the bottom (Figure 5). The weight of the roof created an outward thrust, which the rafters transferred to the walls, tending to push them outwards at the top.

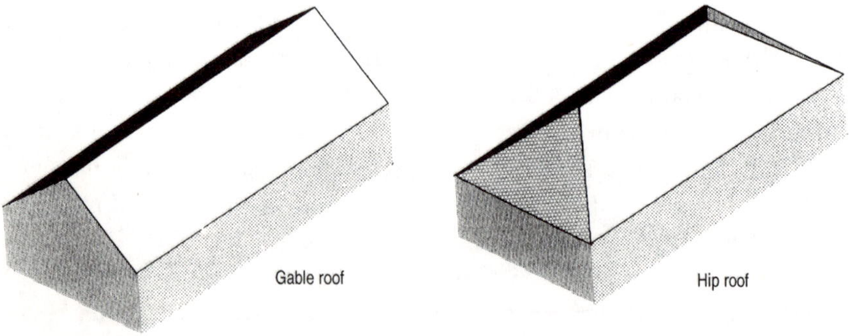


**Figure 4** Hipped roof and gabled roof.
(Drawing by Peter R. Smith)

The problem could have been solved by tying the rafters together with a timber tie, and this was sometimes done in later years. In the twentieth century the ceiling joists were invariably used to tie the rafters together (see p. 33, and Figure 12), but in colonial construction the ceiling joists were generally separate from the roof structure. Another solution, which had been used in English medieval timber structures, was to place a tie across near the top of the roof; this created a near-rigid joint at the crown where the two rafters met. However, in the early years Australian rafters were generally left untied, and this still happened occasionally in the nineteenth century.

Builders were evidently aware of the consequences. The walls were made thick enough to absorb the outward thrust of the rafters, and when the rafters were supported on the columns of a veranda, the columns were often given a slight inclination for the same purpose. More significantly, the spans of colonial rafters were kept low. A span that would be considered perfectly safe with modern tied construction was often halved (Figure 5), and the rafters were supported in the interior on a heavy timber beam. Since roof slopes had to be kept steep with the

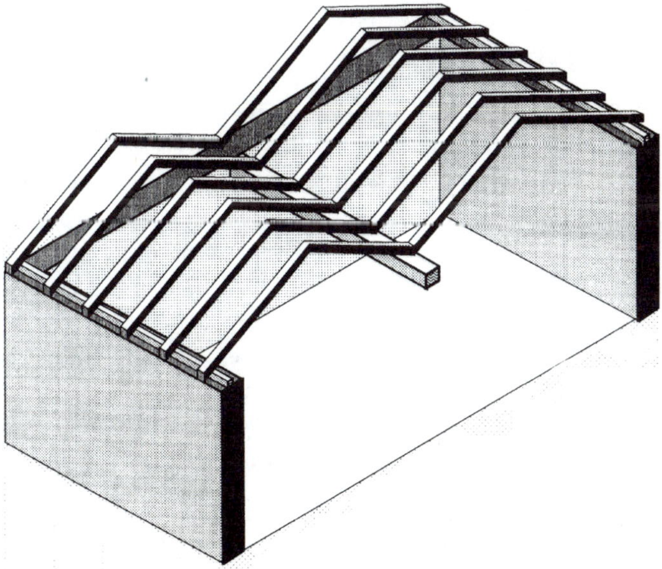

**Figure 5** 'Colonial' roof construction formed by rafters resting on top of the walls without ties. If the span was too great, two sets of rafters were used, supported at mid-span by a heavy timber beam. The roof needed a box gutter to drain the water from the interior slopes.
(Drawing by Peter R. Smith)

early primitive roofing materials, this had the incidental advantage that the roof looked lower and less prominent, but it required a box gutter in the middle of the roof, which frequently leaked.

The roof covering in the early days consisted of rushes, reeds, long grass, and the lance-like leaves of the blackboy (*Xanthorrea*), reinforced with small branches. Thatching was still common in England in the eighteenth century, but as the local materials were unfamiliar, the roofs were sometimes far from weatherproof. They were also highly inflammable, particularly if the building had a chimney, as most buildings did until the twentieth century. In July 1788 Governor Phillip issued an order forbidding the use of thatch on roofs with chimneys.

By that time the brickmakers were turning out clay tiles, but they were also not as weatherproof as their English prototypes. In the meantime, as it had been found that the she-oak (*Casuarina*) could be split cleanly to form shingles, these became a favourite roof covering, until they were banned in urban areas in New South Wales in 1837 by building regulations aimed at fire control. However, Phillip's regulations banning thatch permitted the use of shingles.

## Environmental considerations

Sydney is much closer to the equator than any part of Europe, and the seasonal variation in the length of the day is therefore much less than in London. Artificial lighting would thus not have been a serious concern in the colony's early days.

Furthermore, whale oil for oil lamps was one of the first imports to become readily available. The first whaleships penetrated into the South Pacific a few months after the First Settlement (see p. 110), and Captain Tench wrote that by 1791 he saw four whale ships sailing into Sydney Harbour within six weeks (Tench 1961, p. 298).

Thermal comfort was a bigger problem. The First Fleet arrived in mid-January, and the convicts must have suffered greatly from the unaccustomed heat. Lieutenant Ralph Clark of the Marines made an entry in his Journal every few days 'It is remarkably hott [*sic*]' (Clark 1981, p. 91).

However, the early buildings do not reflect a concern with hot weather. None of the several watercolours and sketches made before 1792 that survive show any houses with sunshading or verandas (see also p. 13). Philip Drew (1992, p. 44) credits Major Francis Grose of the New South Wales Corps, who had served in America, with having built the first Australian veranda in late 1792 or early 1793. Grose arrived in Sydney in 1790, took over the house built by the first Lieutenant-Governor, Major Robert Ross, and added the veranda.

The First Government House (see p. 23) also had no veranda initially; it was probably added by Captain John Hunter after he became Governor in 1795.

There is a difference of opinion (Drew 1992, p. 50; and Herman 1970, p. 19) as to whether the north-facing veranda of Captain John Macarthur's Elizabeth Farm House (see p. 24) is part of the original building, erected in 1795, or a later addition. There is no doubt that the veranda of the Rum Hospital (see p. 28) was part of the original design, because it appears on the contract drawings (Figure 9), which survive; but that was several years later, in 1811.

The veranda, which became such a characteristic feature of the nineteenth century Australian house (see p. 13) was therefore not adopted immediately, even though it would have been familiar to several of the officers, because it was well established in India and in several of the American colonies and former colonies.

The existing pictures of the early houses show another thermal peculiarity: all have chimneys. Many of the chimneys were for kitchen stoves, but others were for fireplaces. Most houses were made of timber, and covered with thatch or wooden shingles. A few early fireplaces were made of brick, but most were built with timber slabs, which were then covered with several layers of clay, and were thus a fire hazard.

It could be argued that heating was not really necessary for the houses near Sydney Cove (see p. 126). Captain Tench kept a thermometer, still a relatively novel instrument, invented by Daniel Fahrenheit in 1716, in his tent and later in his house, and remarked on the rapid variation of temperature in Sydney, but he did not complain about the cold:

> The climate is undoubtedly very desirable to live in. In summer the heat is usually moderated by the sea breeze, which sets in early; and in winter the degree of cold is so slight as to occasion no inconvenience . . . The thermometer has never risen above 84 [°F = 29°C], nor fallen lower than 35 [2°C] . . . Nor is the temperature of the air less healthy than pleasant. Those dreadful putrid fevers by which new countries are so often ravaged, are unknown to us; and excepting a slight diarrhoea, which prevailed soon

after we landed, and was fatal in a very few instances, we are strangers to epidemic diseases. (Tench 1961, p. 69)

People seem to have been used to quite cool indoor temperatures in eighteenth-century England. One of the earliest scientific statements on thermal comfort in winter is an article by Sir Douglas Galton on 'Heating' in the ninth (1880) edition of the *Encyclopaedia Britannica*, in which he recommended a temperature for bedrooms of 'not less than 40 degrees Fahrenheit' (4°C). The temperature in the city of Sydney and its harbourside suburbs rarely falls below this level.

It is therefore strange that so much effort was devoted to providing fireplaces for heating. However, the idea that a house must have a fireplace seems to have been generally accepted, and virtually every house in Sydney had at least one fireplace until World War I.

For a newly established settlement, Sydney had a good water supply. Augustus Alt, the Surveyor-General, had water tanks excavated in the sandstone, near the present Australia Square, to create a storage reservoir in the little stream that ran into Sydney Cove (see Figure 1), and which was thereafter known as the Tank Stream.

Alt also had a wooden bridge built over the Tank Stream near the junction of Pitt and Bridge Streets, to stop people from fording, and thus polluting it. This was replaced by a stone bridge, probably the first stone structure in the colony. However, it collapsed in 1807, and was then rebuilt.

By about 1800 the Tank Stream had become polluted, and Sydney had to rely on several springs for clean water until a generous supply became available with the completion of Busby's Bore in 1837 (see p. 153).

One spring originated near the present Spring Street. It supplied the First Government House, and also a public pump that remained in use until 1905, when it was demolished to make way for the electric trams.

Another was at the western end of the open area that later became Macquarie Place. Governor Macquarie instructed the convict Colonial Architect, Francis Greenway, to erect a stone platform around the pump and a roof over it (Figure 6); it was demolished in the second half of the nineteenth century.

No city anywhere in the world had a complete sewage disposal system in the early nineteenth century. France prided itself on being the

**Figure 6** Morton Herman's reconstruction of the sandstone platform and roof for the fountain in Macquarie Place, built by Francis Greenway, the first Australian ornamental sandstone construction. It was an important part of the water supply after the Tank Stream became polluted about 1800, and before Busby built the water supply tunnel in 1837.

(Reproduced from Herman [1970, p. 35], by permission of Harper Collins)

most civilised country at that time, but it conceded that Britain was the inventor of the water closet (WC), which was known as a *cabinet d'aisance à l'anglaise.* These were common in upper-class houses in England in the early nineteenth century, and it may be that one was installed in Government House during Governor Hunter's term of office (1795–1800). A sewer drain was certainly built under Government House around 1828, when Ralph Darling was Governor (Proudfoot et al. 1991, p. 52). However, installation of a drain did not necessarily constitute sewage disposal as we understand it today. The drain would have discharged into a soaking pit or into Sydney Cove. In a country house, which the First Government House originally was, this was reasonably satisfactory. The building was demolished in 1845 (see p. 23), and even in London there was no general sewage disposal system until 1856.

## The First Government House

Governor Phillip had brought with him a prefabricated house made of canvas and timber, for which the Admiralty had paid £115. It was manufactured by Samuel Wyatt, an Uncle of Jeffry Wyatt, who later as Sir Jeffry Wyatville became the architect for the restoration of Windsor Castle, and who has been credited with the design of the prefab (Cox &

Freeland 1969, p. 16; Herman 1970, p. 8). Construction of this house started during January 1788 west of the Tank Stream overlooking Long Bay (later called Darling Harbour). The Natural History Museum in London has a watercolour of the erected building, which is reproduced by Proudfoot et al. (1991, p. 32). The house proved far from waterproof in heavy rainstorms in February, and planning for a more weather-resistant building on a site east of the Tank Stream started in March. The foundation stone was laid on 15 May 1788, and Governor Phillip moved in on 4 June 1789.

Several explanations have been offered for the move of the Government House across the Tank Stream. One is that Phillip did not like his Lieutenant-Governor Major Ross, and wished to be some distance from the house he was already building; another is that he preferred the splendid view from the new site. The most likely is that initially this building was also considered temporary. The map drawn by Lieutenant William Dawes in July 1788 (Figure 1) gives as references: '1. A small House building for the Governeur' (i.e. the First Government House), and '5. Ground intended for the Governeur's House' (a site on the same side of the Tank Stream as most of the First Settlement).

The original intention was to build just three rooms (Figure 7(a)). However, when the walls went up Phillip accepted the advice that the brickwork was strong enough to support a second storey, and the design was revised. It thus became the first house in the colony with a staircase, and remained the only one for some time.

Alt's bridge over the Tank Stream (see p. 18) was in line with the house, and the present Bridge Street is over the path that led to it. Phillip Street is over the path from the house to the Cove. When the demolition of the First Government House was completed in 1846, Bridge Street and Phillip Street were extended over the site, so that the greater part of the foundations of the original house now lie under their pavements.

Some information was obtained during road repairs in 1899, but the archaeological investigation of the foundations in 1983 and 1984 had to be restricted to that part of the foundations that lay beyond the pavements. Furthermore, the investigation was limited to the foundations and what could be found below ground, because nothing remained of the superstructure (Proudfoot et al. 1991, p. 1).

The builders of the original house had removed the soil, and placed the foundations directly on the solid rock. They were 2 feet 6 inches

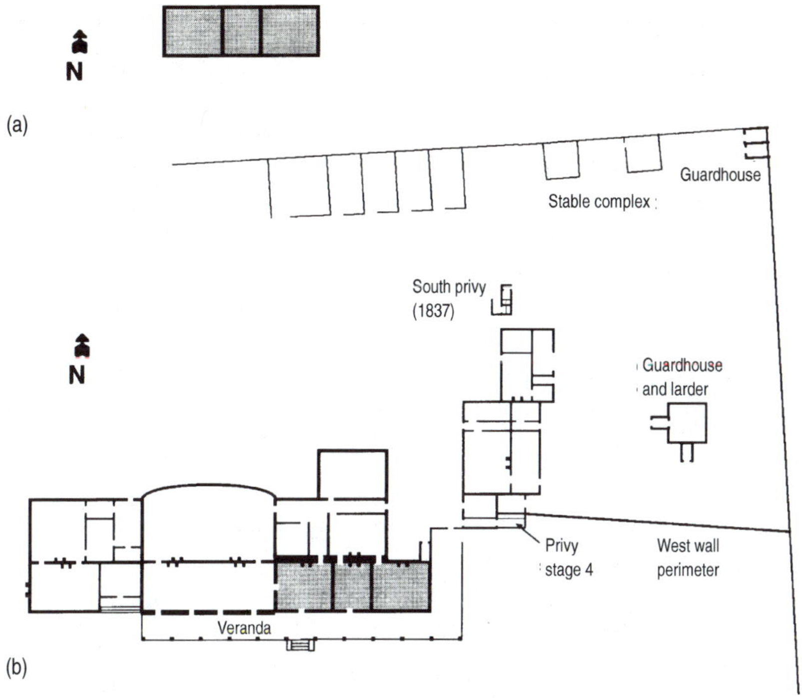


**Figure 7** Plan of the First Government House: (a) the original house built by Governor Phillip; and (b) the final form before demolition. The north-facing curved facade on the main building was added by Governor Macquarie.

(Drawing by Peter R. Smith from plans published in Proudfoot et al. [1991, pp. 26–7], made by permission of Allen & Unwin)

(0.76 m) wide and constructed in two rows of sandstone blocks dressed on the outside, but irregular on the inside. Broken bricks and mortar (which was pure pipe clay dug from the surface of the ground on the site) were packed between the stones. They thus formed a solid foundation.

In a few places the remains of the brick walls were found. These were made of salmon-coloured bricks mortared in pipeclay. This fine white clay contained no lime. Anne Bickford, who led the excavation, said that the bricks were similar in colour and in size to those she had previously excavated at Elizabeth Farm House, built in 1795 (see p. 24). Some of the materials recovered during the excavation were subsequently analysed by the Australian National University's consulting company to establish its composition and origin (Pearson 1988).

A number of officers and convicts sketched or painted the original Government House, and Proudfoot (1991, p. 32) reproduced one of these. The house was built of brick; the thickness of the wall was probably slightly less than that of the foundation (0.76 m), with quoins of sandstone.

Morton Herman (1970, p. 11) states that the bricks were originally laid in mortar of puddled clay, which was the usual mortar in the early days of the Colony, but that part of a wall collapsed during a rainstorm, and the walls were then rebuilt with lime mortar made from seashells, the only building for which lime mortar was used in Sydney in the eighteenth century. He thought that the use of lime mortar made it possible to add a second storey. A lime kiln had been built at Bennelong Point, burning the seashells that were being collected.

It is assumed that the building was originally plastered with clay, and finished with a coat of white pipeclay, but was later replastered with lime stucco and whitewashed. The windows had crown glass brought by the First Fleet (see p. 3).

It is further assumed that the house was originally covered with clay tiles, pieces of which have been found in the foundations, but as these showed early signs of decay they were replaced with she-oak shingles, which have also been found in pits on the site.

Most of the people who wrote diaries in the early days of the colony commented on the severity of the thunder and lightning as compared to England, and in 1790, about one year after completion, a lightning conductor was installed on the roof of the First Government House after a bad thunderstorm. The house stood on high ground, and it was the only two-storey building. The fact that an iron rod at the highest point of a roof and connected to the ground by metal would protect a timber roof from a lightning strike, which might set it on fire, was discovered by Benjamin Franklin in 1750, while he was living in England. It was thus still a relatively novel concept. The introduction of the lightning conductor was important for a country where timber structures continued to play a far more important part than in Europe (see p. 35).

There is no doubt that the house was built by James Bloodsworth, a master bricklayer, who arrived with the First Fleet in 1788, having been sentenced to seven years' transportation. He was pardoned in 1790, and in 1791 was appointed superintendent of all bricklayers and brickmakers.

He is also generally credited with the design, but Helen Proudfoot (1991, p. 55) thinks it may have been the work of Henry Brewer (see p. 4), an old friend of Governor Phillip, who had some architectural training, and was the unofficial superintendent of construction in the colony during Phillip's term of office. The building had a well-proportioned facade; its central bay was strongly defined by a pediment breaking the roof line, and by a fanlight over and sidelights around the front door.

Most subsequent governors made additions. Hunter probably added the veranda (see p. 17). Macquarie added the curved northern facade (Figure 7 (b)), to accommodate a room which he described to Lord Bathurst, the Secretary for the Colonies, as an Office, but in his own Memoranda Book as the Great Saloon (Broadbent 1992, p. 160).*

The porosity of the brickwork of the original building caused dampness in the walls, which drew adverse comments from Lachlan Macquarie and all subsequent governors. In 1832 Richard Bourke requested a plan for a new Government House 'by an eminent architect in London'. Edward Blore's Neo-Gothic design was accepted in 1837, and construction of the present New South Wales Government House started in 1838 under the supervision of the Colonial Architect, Mortimer Lewis. Blore never came to Australia.

In 1845 Lewis recommended that the old building be demolished, because its repair would cost more than a new building. Proudfoot (1991, p. 129) comments that the *ad hoc* nature of the First Government House would have been an irritation to an architect who favoured a formal style (see p. 44). It is a measure of the change that had occurred in the half-century since its foundation stone had been laid.

In 1995 the Museum of Sydney, located just north of the site, was opened with exhibits of the First Government House. Part of its foundation is displayed under glass in its forecourt and foyer.

## Some other early buildings

The oldest surviving house in Australia is Elizabeth Farm House, built by Captain John Macarthur of the New South Wales Corps. It was a farm house in Parramatta (originally called Rose Hill after the Secretary of the Treasury), where food production was concentrated in the first few years after the failure of the crops planted near Sydney Cove.

* Francis Greenway, who became the Colonial Architect after the room was built, gave evidence at the inquiry by Commissioner Bigge into alleged extravagance by the Governor. He said that Mrs Elizabeth Macquarie had brought with her a new architectural pattern book by the English architect Gyfford, and that the room was based on Gyfford's 'Tenth Design' (Gyfford 1806, p. 19, and plates XXV and XXVI).

The date is variously given as 1793 and 1795, and it has been illustrated in several books (for example, Freeland 1968; Baker 1961; Drew 1992). It is a single-storey brick house, planned and constructed similar to the First Government House, except that no lime mortar was available for the brickwork. The walls would have been rendered originally with clay, and the present coat of stucco is more recent. Drew considers that the stone-paved veranda on its north side was a later addition, but others think it was part of the original building. Its roof would have been originally supported on wooden posts; the present cast-iron columns (see p. 63–64) date from the 1860s. The house was altered and added to several times.

An early building that survives almost unaltered is Experiment Farm Cottage, also in Parramatta. It is illustrated by Baker (1961) and Drew (1992), and dates of erection have been given between 1798 and 1820. It is similar to Elizabeth Farm, except that it has a wide veranda, which was originally supported on timber columns, completely surrounding the house. The house was not built by the original owner of Experiment Farm, James Ruse, the first person in the colony to grow enough food to feed himself, but by John Harris, who bought the farm from him.

The 'Old' Government House, which is still standing in Parramatta Park, was the residence of the Governors in Parramatta (Proudfoot 1971); in the early years of the colony the journey from Sydney Cove to Parramatta took at least one day, sometimes two, by boat, and longer overland. Governor Phillip built a small single-storey house in Parramatta in 1790, and it may be (O'Brien 1993, p. 10; Mackay 1995, p. 13) that this was the core of the present Old Government House, and that Governor Hunter merely added a second storey in 1799. The old part of the house is built with plastered brickwork of poor quality, but the extensions by later governors, also of brick, are well built. The Tuscan-columned portico was part of the extension in Governor Macquarie's time.

The house was a second residence, except in the time of Governor Brisbane, who spent more time there than in Sydney. In 1848 a carriage driven by Governor FitzRoy overturned a few hundred metres from the house. His wife was killed, and thereafter neither he nor any other governor lived in it. It was restored and reopened in 1970.

It is a modest building compared to the great English country houses of the period, but it has a pleasant setting in a garden surrounded by a park, which was also the setting of the First Government House in its early days.

The Commissariat Store (illustrated by Herman 1970, p. 27), and by Balint (1984, p. 13), built on the site now occupied by the Museum of Contemporary Art, belongs chronologically to this period, but its building technology was more advanced than that of the other buildings discussed. It was designed by Lieutenant-Colonel Joseph Foveaux; the plans, dated 20 February 1809, are in the Mitchell Library (Herman 1970, p. 28). Foveaux came to Sydney as a Lieutenant in the New South Wales Corps in 1789, returned to Britain in 1811, and retired from the army with the rank of Major-General. It is not known whether he specified the internal construction. Governor Macquarie, in a despatch to Lord Liverpool in November 1812 wrote: 'Col. Foveaux commenced a granary and provision store on a large scale. I continued the work until it was finally complete, little more than the foundation having been laid at the time of my arrival' (Balint 1984, p. 12).

Since the building survived until 1938, when it was demolished to make room for the Maritime Services Building (now the Museum of Contemporary Art), its outside appearance is known from illustrations made at various times (see Herman 1970, p. 27). It was four storeys high, and the external walls were of sandstone. Burglary of government stores was a major problem in the early days, and it would have been more difficult to break into a multi-storey building whose walls were made from large stone blocks, than into a single-storey building of brick and timber.

Balint (1984, p. 12) assumed that the interior structure consisted of hardwood storey posts, girders and floor planks, thus becoming the prototype of the later warehouse and store buildings (see p. 37).

Not all the early buildings were purely utilitarian. James Bloodsworth built a brick clock tower on Governor Hunter's instructions near the site where St Philip's Church was built later (York and Jamison Streets). The church clock had arrived in the *Reliance* in 1795. The tower can be seen in several contemporary sketches (for example, Figure 8), and it appears in Hunter's works record for 1797. Herman (1970) and Bridges (1995) give the height of the tower as 150 feet (46 m), but Figure 8 shows a tower that looks more like 50 feet in height.

The tower was damaged in a storm in 1799, and collapsed in 1806 (Bridges 1995, p. 37). Herman (1970, p. 21) blamed the failure on the

**Figure 8** The General Hospital at Dawes Point, which functioned from 1796 to 1816. Governor Hunter's clock tower, built in 1797, is seen in the background in the middle of the picture.
(Drawing by Peter R. Smith, based on a sketch by an unidentified contemporary artist, reproduced by Watson [1911, p. 7])

low strength of the bricks, but the poor quality of the mortar in 1797 and an inadequate thickness of the walls for stability under wind load were probably more serious defects (see Cowan 1977, p. 110).

## Early hospitals

The first hospital was established immediately on arrival in Sydney Cove in a number of tents. Governor Phillips instructed Henry Brewer in February 1788 to erect a wooden hut for the hospital to supplement the tents. It was probably near the present Museum of Contemporary Art.

In June 1790 the Second Fleet brought a large prefabricated wooden hut, made like Phillip's temporary Government House by Samuel Wyatt (see pp. 19 and 37), but with a copper roof covering (Bridges 1995, p. 16). Added to the locally made wooden huts (Figure 8), the hospital now had room for sixty beds (Watson 1911, p. 3). In 1796 the huts were re-erected nearby on sandstone foundations.

Governor Lachlan Macquarie arrived in December 1809, and on 8 March 1810 he wrote to the Secretary for the Colonies, Viscount Castlereagh, that 'there will be an absolute necessity for building a new general hospital as soon as possible, the present one being in a most ruinous state' (Watson 1911, p. 15).

He did not wait for a response, but instructed the newly appointed Acting Principal Surgeon, D'Arcy Wentworth, to advertise in the official *Gazette* on 19 and 26 May 'that a Plan of the above [hospital] will be ready at my house for inspection of those desirous of contracting for the same.'

The drawings, which consist of a large plan and a large elevation of the buildings, are reproduced by Watson (1911, pp. 20 and 32). They are not working drawings, but they are properly scaled and lettered, the work of an experienced draughtsman. Commissioner Bigge was unable to ascertain who designed the hospital, and that may be because the Governor himself provided the basic design, before the appointment of the first Colonial Architect. The buildings are surrounded by colonnaded verandas on two floors (Figure 9), and they bear a distinct resemblance to those of the Government House in Madras (Fletcher 1987, p. 1266), which would have been well known to Macquarie from his service in India.

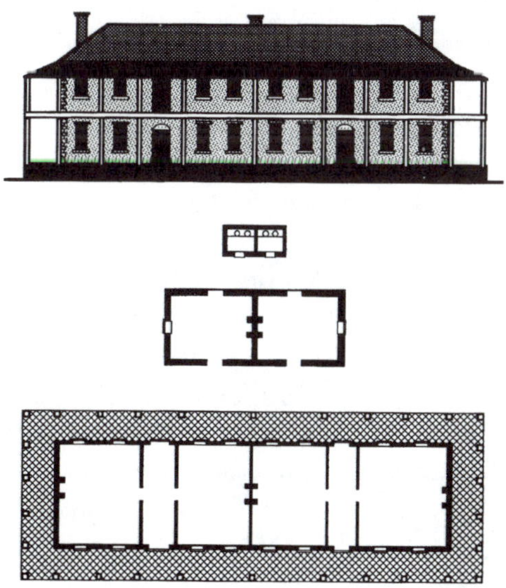

**Figure 9** Plan and elevation of the Stores Building of the General Hospital, which later became the Mint.

(Drawn by Peter R. Smith from the 1810 contract drawings, reproduced from Watson [1911, pp. 20 and 32])

The hospital was a two-storeyed building 288 feet (88 m) long. In addition there were two smaller buildings of similar design on either side, one for accommodating the surgeons, and one for stores.

As is well known (for example, Watson 1911, p. 23; Herman 1970, p. 33; Hughes 1987, p. 297) the 'Rum Hospital' was financed by giving the contractors a three-year monopoly of the sale of rum, a six-month delay in the payment of the excise, the free labour of twenty convicts, the use of twenty draught bullocks, and eighty oxen from the government herd for slaughter. Opinions differ on whether the contractors made a modest or a big profit, but the colony got a large hospital at a time when it did not have the money to pay for it.

The main hospital building was demolished in 1881 to make room for the new Sydney Hospital. The Surgeons' Building, never fully occupied as such, was partly requisitioned in 1829 by Governor Ralph Darling for the Legislative Council, and greatly altered after the Legislative Council was enlarged.

The Store Building became the Royal Mint in 1865. In the 1970s it was restored as far as possible to its original state, and became the Mint Museum.

This building had plastered walls 2 feet 3 inches (0.7 m) thick. The lower veranda was paved with flagstones that came from Cockle Bay (Darling Harbour). The Tuscan columns supporting the upper veranda are of sandstone quarried in the present Domain. The columns were subsequently rendered, probably because the finish was poor. Herman (1970, p. 33) comments that the columns bear little relation to those of classical architecture, but neither did those of the Madras Government House. The columns of the upper storey of the Mint Building supporting the roof were of hardwood, mostly Sydney blue gum and mahogany.

The building has a colonial-style roof in two parts, with a central gutter (see p. 15), but it has ties. However, the ties are in three parts. The joints are supported by the walls, and are only connected by small plates. Francis Greenway, who became Colonial Architect while the hospital was under construction, was instructed to investigate complaints made after its completion. He severely criticised the inadequacy of the joints. It may well be that the builders were aware of the problem, and deliberately built the upper columns leaning slightly inwards to absorb any outward thrust of the roof (see p. 15).

Greenway made a long list of constructional defects, and predicted that the buildings would soon collapse. However, this did not happen, probably because the walls were thicker and the timbers bigger than was strictly necessary.

The design has been criticised not merely by Colonial Architect Greenway and by Commissioner Bigge, but by architectural historians. However, the Mint is an interesting and charming building, which can hold its own in a street that has some good old and modern architecture. The three buildings of the Rum Hospital, which presented a two-storey frontage of almost 200 metres, were the largest project so far undertaken in Australia, and also among the last important buildings employing primitive construction .

# 2

# TRADITIONAL CONSTRUCTION

## Domestic timber construction

The story of building construction during the first few years of the new colony has been related in some detail, because it is unique. Settlements had been founded before by colonists or adventurers who went of their own choice. Convicts had been sent to established colonies, which had an existing organisation. But the 'policy of settling, with convicts only, a country at once so remote and so extensive', as Captain Watkin Tench put it so perceptively in 1793 (see p. 6), was new, and it created its own problems.

These were gradually overcome. It had been generally assumed that the Napoleonic Wars had ended in 1814, and this freed shipping and other resources. In fact, Napoleon made a come-back in 1815, but this did not last long. Free immigrants had been coming since 1793, and after 1815 they came in increasing numbers.

The crossing of the Blue Mountains in 1813 gave access to vast new resources, including limestone and new species of timber.

By the mid-century Australian building construction did not differ greatly in technique and quality from British building construction, whose principles can be studied from nineteenth-century textbooks. The main difference was in the use of timber, which was partly due to the peculiarities, both desirable and undesirable, of the Australian species, and partly to the influence of American innovations in light timber framing.

Log cabins, which figured prominently in the early history of American building construction (Figure 10), were less common in Australia. Instead the horizontal timbers were fitted into vertical posts, which had been grooved, or after nails became more freely available, nailed to them. An alternative method was to use vertical slabs (Figure 11). The timber was either hand-sawn, or split with a maul and wedges, and then finished with an adze. After power saws became available in the 1830s these methods of construction, particularly that in Figure 11(b), became much cheaper and more weathertight.

A more durable construction based on the wattle-and-daub concept was also widely used in the early nineteenth century. Cox and Freeland described it:

> A sturdy frame of heavy timber was erected, nogging panels of brickwork or stone were set between heavy posts in a kind of half-timbering. The outer face of the walls was sheeted horizontally with weather boards about 6 inches [150 mm] wide and 1 inch [25 mm] thick in profile, split from box, ironbark, stringy bark or she-oak. The result was a structurally sound, completely weathertight and well-insulated building. The inner face of the brickwork could be plastered or painted, and the buildings attained a degree of quality that made them acceptable. With the aid of iron knees and end plates two-storey buildings were erected in which the upper storey

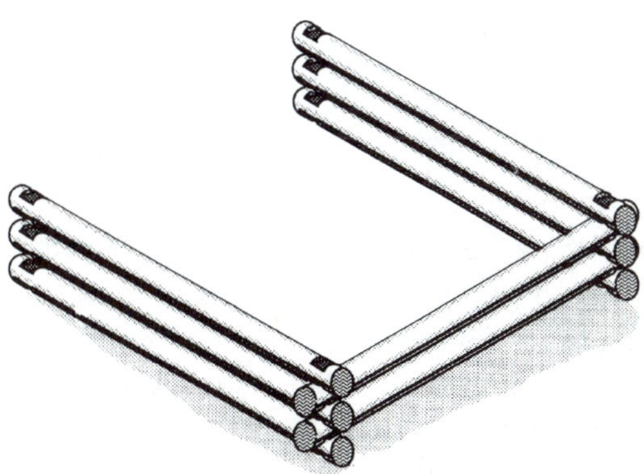

**Figure 10** Log cabin, built from whole round timbers. These were halved near the ends to form strong right-angle connections.
(Drawing by Peter R. Smith)

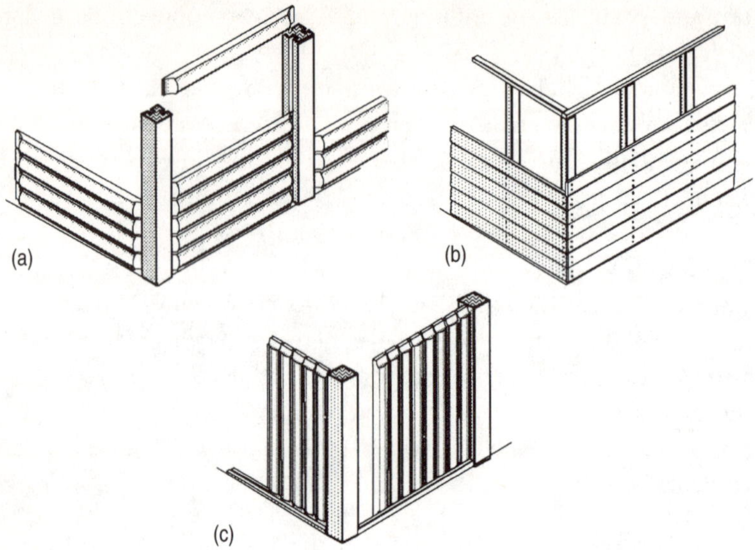


**Figure 11** Timber slab buildings: (a) horizontal slabs fitted between vertical posts; (b) horizontal slabs nailed over vertical posts; (c) vertical slabs.
(Drawings by Peter R. Smith)

> of brickwork was carried on heavy timber beams independently of the lower walls . . . Many of them still stand in excellent condition after 140 years or more, and one that has been claimed to be the first of them and which, in any case, is certainly one of the earliest, Rose Cottage at Wilberforce erected in 1798, is in daily use. (Cox & Freeland 1969, p. 34)

Since the brickwork is inside, this construction is, in a sense, a reversal of the half-timbered houses that have been built in England and many parts of northern Europe since the Middle Ages, and were still being built in the nineteenth century. There are a few of these in Australia, but they are mostly in the part of South Australia that was settled by German migrants after the 1848 Revolution.

The light timber frame, which is now the most widely used structure for Australian single-family houses (Figure 12), probably came to Australia in the 1850s after the gold rush. It bears a strong resemblance to the balloon frame, which was first used in a Chicago church in 1833. It may have been brought back by Australians who went to California in 1849 to dig for gold, and returned when large quantities of gold were discovered in Victoria. In modern light timber construction the rafters are tied by the ceiling joists, and lateral stability is ensured by diagonal bracing (Figure 12). The outer walls can be weatherboard, or brick or any weatherproof material. The inner walls are of light construction. In the original 'Queensland style', discussed below, they were often omitted.

These houses were originally made entirely of timber, but weatherboards are now rarely used except in Queensland and Tasmania. The brick-veneer house, which is one of the original Australian contributions to building construction, is a house with a timber frame, clad externally with a single leaf of brick. It looks like a brick house, but its structure is a timber frame.

Ray Sumner (1985) considers that the weatherboard houses of the 'Queensland style' are an independent development. Queensland was settled later than the other states, and free settlement was permitted in Brisbane only in 1841, so that saw mills in Queensland were steam-powered almost from the beginning. It was thus possible to produce much smaller timber sections capable of carrying the load safely than would have been possible by hand-sawing or splitting. Unlike the American balloon frame and the present standard Australian light timber frame, which are only skew-nailed, the framing of the early

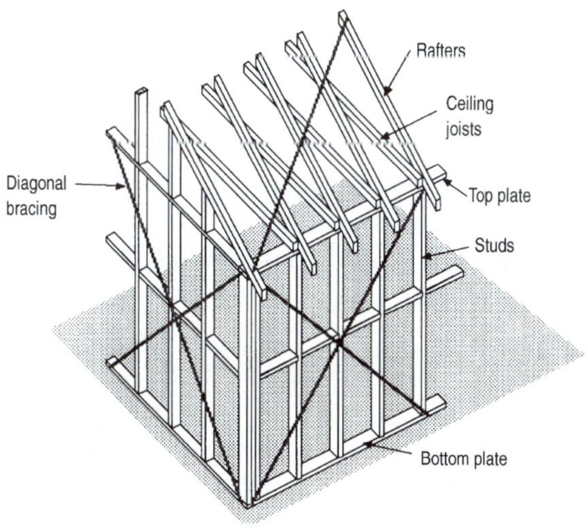


**Figure 12** The light timber frame: the rafters are tied to the ceiling joists, and lateral stability is ensured by diagonal bracing.
(Drawing by Peter R. Smith)

Queensland houses had studs that were tenoned, and were skew-nailed as well. The sheathing was initially with a single layer of weatherboards, usually on the outside only, but occasionally on the inside only.

The Queensland style is generally assumed to include elevation on storey-high stumps; but Sumner points out that only about one-third of the Queensland houses in the later nineteenth and early twentieth centuries were elevated.

The concept of putting the houses on hardwood stumps had its origin in the short-lived unsuccessful settlement of Port Essington (near present-day Darwin) in 1839, where the termites were far more voracious than in Sydney, so that the time available to take preventive action was much shorter. If the house was put on tall stumps of a hardwood that was termite resistant, the white ants had to build easily visible tunnels on the outside of the stumps.

Putting houses on tall stumps had other advantages. It created useful space under the house, which was thermally insulated by the house above it; and in the warm-humid region it permitted ventilation under the house. On the other hand, it made the house more susceptible to wind damage during cyclones.

The metal caps over the brick stumps, on which the termite-susceptible timber structure was supported, came into use later in the century. They are now bent down and soldered to form a cap, so that the termites would have to build tunnels down and outwards to get past. However, the correct design took some time to develop, and failures were still recorded in the 1950s.

The introduction of power saws solved the problem of cutting Australian hardwood, but nailing was still hard work. For this reason in the first half of the nineteenth-century European spruce was imported from England, probably grown in Scandinavia or Russia. American softwood, particularly Douglas fir, generally called 'oregon', was imported after the development of a timber industry on the American West Coast. The planting of Monterey pine (*Pinus radiata*), a native of California, started in Australia in the 1880s, but it gained momentum only during World War I. Due to the shortage of softwood during World War II some of these trees were used before their maturity, and the acreage of radiata forests was greatly increased.

The first free settlers arrived in Western Australia in 1829. They were far better informed about the land, its resources and its problems than the people of the First Fleet sent to Botany Bay in 1788; but they had some surprises, not all unpleasant. They discovered a plentiful supply of two trees, jarrah (*Eucalyptus marginata*) and karri (*Eucalyptus diversicolor*), which did not grow in the eastern colonies, and whose timber was as durable as that of ironbark, but easier to work.

## The fire problem

The main structural problem of traditional construction had remained unaltered until the mid-eighteenth century. Timber could be used for floor and roof structures, as well as for the walls and columns, but it could be destroyed by fire. Stone and brick were durable and fire-resistant materials, and were excellent for walls and columns, but to span a horizontal gap one needed to use them in the form of arches, vaults or domes (see p. 54). In the fifteenth century Alberti put the case for them:

> I am entirely for having the roofs of temples arched, as well because it gives them greater dignity, as because it makes them more durable. And indeed I know not how it happens that we shall hardly meet any temple whatsoever that has not fallen into the calamity of fire . . . Caesar owned that Alexandria escaped being burned, when he himself took it, because its roofs were vaulted. (Alberti 1955, p. 150)

Lightning was a particular cause of fire in medieval and Renaissance cathedrals, because lightning is liable to strike the tallest object in the vicinity. However, this ceased to be a problem after the invention of the lightning conductor in 1750 (see p. 22), and nearly all Australian cathedrals have timber roofs without interior stone vaulting (see p. 55).

War and civil war have caused, and are still causing a great deal of destruction by fire in Europe and Asia, and America has not been entirely immune; but few people in Australia have ever been seriously concerned about designing buildings to resist fire caused by war, except briefly during World War II (McMillan 1944, p. 43).

Moreover, some Australian timbers are highly fire resistant. Ironbark (in the east) and jarrah (in the west) can survive an ordinary warehouse or office fire if the cross-section is big enough. The fire chars the timber to a thickness of about 25 mm, and this serves as insulation for the rest of the section, which retains its structural integrity. The structure will not collapse if the structural sections were somewhat oversized before the fire.

The 'fireproof' construction developed in Britain in the later eighteenth and the early nineteenth century gained only partial acceptance in Australia.

The Industrial Revolution, which started in England in the eighteenth century, created a new fire problem, because of the assembly of combustible materials, such as cotton and wool, and of a large number of people in multi-storey buildings. The important innovation was the use of a hydraulic or steam engine to substitute mechanical for muscle power. The textile machines were placed as close as possible to this prime mover to reduce the loss of power during transmission; hence the use of buildings four or five storeys high to shorten the length of the belt drives.

The first step was to replace the timber floors with brick jack arches in the 1790s (see Cowan 1977, p. 249, fig. 8.11). In 1811 William Strutt substituted cast-iron for the timber beams and columns. Cast-iron beams and columns were used in Australia from the 1840s (see pp.62 and 70), but brick jack arches never became popular here, and the few examples of their use were to be found mainly in railway bridges (for example, on the Sydney to Parramatta line).

Because of the high fire-resistance of hardwoods, such as ironbark, the Commissariat Store (see p.25) became the prototype of a form of 'colonial' construction (Figure 13). It survives in a substantial number of buildings, particularly in Hobart (Figure 20), and to a lesser extent in Sydney. These have square storey posts and rectangular beams of a timber such as ironbark, essentially the construction that was used in England until the end of the eighteenth century, except that the Australian timber was much more fire resistant. The floor was sometimes made of hardwood joists and boards, and sometimes of more nailable, but also more combustible, softwood. Terra cotta lumber (see p. 70) was used in conjunction with wooden floors from the late nineteenth century onwards.

In the second half of the nineteenth and the early twentieth centuries iron (and later steel) columns and beams were used more and more, but hardwood structures continued in use well into the 1920s, when reinforced concrete was generally adopted for multi-storey floor construction (see p. 100).

There is one late example. The Goldsbrough Mort Woolstore at Pyrmont in Sydney had been built in 1882 with cast-iron columns and wrought-iron beams, which were not given fire protection. They were

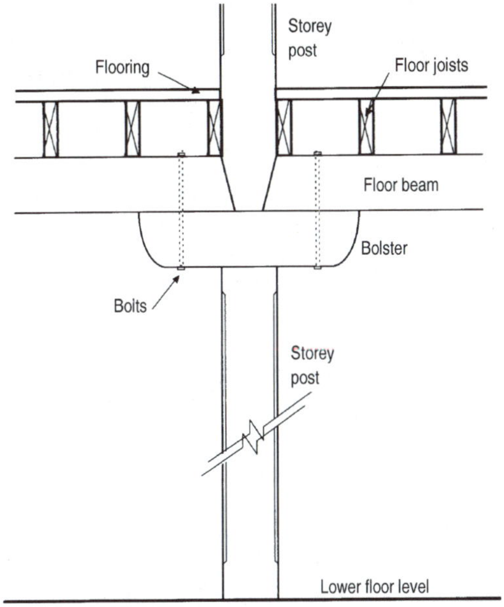


**Figure 13** The use of ironbark storey posts and beams in warehouse construction: the floor consists of joists of imported Douglas fir and boards of hoop pine, one of the few Australian pine woods (Commonwealth Wool and Produce Store, Jones Street, Pyrmont, NSW, built in 1899).
(Drawing by Peter R. Smith, based on Balint [1984, p. 103])

thus incombustible, but not fireproof. When the store was totally destroyed by fire in 1935 it was decided to revert to traditional construction, and it was rebuilt in 1936 with ironbark storey-posts and girders. This was their last use in Sydney.

## Prefabrication

As mentioned in chapter 1, the prefabricated house that Governor Phillip brought with him in 1788 was not in use for long. The prefabricated hospital, which arrived with the Second Fleet in 1790, however, lasted for many years. It was one of a series built 'for His Majesty's distant possessions' by Samuel Wyatt, a London builder, whose clients included the engineering firm of Boulton & Watt (Ware 1967, p. 164). Wyatt wrote to Boulton: 'I exhibited the movable Hospitals to the King &c &c by taking down one of the buildings & putting it up again . . . in one hour, which gave general satisfaction' (Herbert 1978, p. 5).

In 1804 Governor King ordered some wooden huts to be made, ready-cut for easy erection in the new penal settlements of Newcastle and Hobart.

When Charles La Trobe arrived in Melbourne in 1839 as Superintendent of the new settlement, he brought with him a prefabricated wooden cottage made by John Manning in London. This served as Government House until 1854. It has now been re-erected in the Domain, but judging by the appearance of the building before the move from Jolimont (Casey et al. 1963, p. 128) there has been substantial restoration (National Trust 1987).

In 1837 Manning started to advertise in the *South Australian Record*, a newspaper published in London for intending immigrants, and by 1840 he included a list of testimonials from satisfied customers in Adelaide and in Perth.

The settlement of Perth in 1829 and of Adelaide in 1836 by free settlers was greatly facilitated by their bringing prefabricated cottages, although only some of the settlers had the means to purchase and transport them. All these houses were made of timber; none survive.

In 1851 Victoria became a separate colony, and shortly thereafter the discovery of gold in paying quantities caused a huge influx of population, which increased from 76 000 before the gold rush to more than half a million by 1860. Victoria overtook New South Wales both in population and in wealth. This resulted in the importation of a large number of prefabricated buildings, not merely for housing, but also for hotels, theatres, and churches. Most were made of iron, and these are discussed in chapter 3.

## Brick construction

During the Roman Empire both hard-burnt brick (usually just called brick) and natural stone were used for masonry. However, in Ancient Egypt and Greece only natural stone and sun-dried bricks had been in use, and in Mesopotamia there were no deposits of natural stone. After the fall of the Roman Empire the art of brickmaking declined in Western Europe. Bricks became more expensive than stone, and were used only in regions without natural stone, such as the Low Countries, North Germany, and the East of England.

In the fifteenth century brick again became more common as technical innovations in brick manufacture were introduced, and these were brought to England by Flemish brickmakers. The revival of brick masonry was therefore still relatively new in the eighteenth century.

Alec Clifton-Taylor wrote about Georgian England:

> To explain the triumph of brick is not difficult. It was a question of economics. Long before bricks were made by machinery this was so. Provided the right clay was available, bricks were not expensive to make. Although time and labour and reasonable skill were involved, the cost was much less than that of quarrying, shaping and transporting, even for short distances, blocks of heavy stone. (Clifton-Taylor 1962, p. 62)

He also put the case for stone:

> But why, it may be asked did anyone trouble to face his house with stone, especially in 1700 when bricks of the most excellent quality were becoming widely available? The explanation is a social one. For throughout the whole of the eighteenth century, and well into the nineteenth, stone carried with it a social cachet which attached to no other material. The Georgians, although excelling as builders of brick, had a yearning for stone beyond anything that had been felt in this country hitherto. For monumental public buildings, stone was considered well-nigh essential; for houses, although not essential, it was certainly highly desirable. (Clifton-Taylor 1962, p. 62)

It was immediately evident when the First Fleet arrived that Sydney had a good supply of sandstone; but it also had a supply of suitable clay at Brickfield Hill and a handy supply of wood-fuel and water (see p. 12). By the 1830s, when conditions in the colony had become somewhat easier, stone replaced brick as the material for prestigious buildings in Sydney. This, however, did not apply to some other Australian cities where there was no local natural stone of good quality.

In the early years bricks seems to have been generally used with a coat of plaster (See pp. 14, 22 and 51). By 1815, when Francis Greenway became the first Colonial Architect, the quality of the bricks had improved, and he used face bricks in most buildings, with very satisfying results (Figure 14).

In Sydney brick buildings were sometimes erected with stone quoins, or even complete stone corners, a practice introduced from England. In Adelaide, where the local building stone was not of high quality, a reversal of this practice developed into an Early Adelaide Vernacular (Morgan & Gilbert 1969, p. viii): the bricks formed the quoins or corners of the

**Figure 14** Hyde Park Barracks, Sydney, designed in 1817 by Francis Greenway, with brickwork of a quality unattained in the previous decade. The arches are well-formed, and specially selected hard bricks were used for the lower level to provide a moisture barrier, at a time before the introduction of damp-proof courses, The roof is not 'colonial', but has kingpost trusses spanning 13.2 m.
(Photograph by Henry J. Cowan)

building, and the body of the wall was built from local stone, in the form of random or squared rubble (Figure 15). A limestone, which in many parts of Adelaide is just below the surface, was mainly used, or alternatively a 'bluestone', which is actually a form of slate.

The same construction was used in early Melbourne buildings with the local bluestone, which is a hard basalt. There are also a few examples in Perth with limestone.

Because of the absence of good local natural stone, brick was used for a number of important public buildings, such as the Anglican cathedrals in Newcastle (NSW) (by Horbury Hunt, started in 1869), in Armidale (NSW) (also by Hunt, started in 1871), and in Perth (by Edmund Blacket, started in 1880).

The Independent Church in Melbourne, built by Reed and Barnes in 1866–7, is a good example of the use of polychrome brickwork in geometric patterns (Figure 16) in the then-current English fashion, and the range of colours shows how far Australian brick manufacture had advanced by then.

**Figure 15** Building in North Adelaide with rubble walls edged in brick, and brick chimneys. (Photograph by Henry J. Cowan)

The manufacture of bricks was initially a government operation, using mainly convict labour. By 1837, however, a number of small brickyards had been established by private enterprise. Furthermore, small brick ovens were sometimes set up at homesteads, as it was simpler than transporting bricks over bad roads.

By 1870 New South Wales brickworks were being equipped with steam-driven moulding machines (Gemmel 1986, p. 6). The Platt dry-press machine, which could push out four bricks at a time, was widely used.

By that time Melbourne had overtaken Sydney as the fastest-growing city, with the greatest demand for bricks. In 1870 the Hoffmann Patent Steam Brick Company was opened in Melbourne, using a wholly mechanised and continuous German process. However, the small brickworks continued to function for some time, and James Nangle, who later became the Head of the Building Department of the Sydney Technical College, in 1894 thought that the handmoulding operators were turning out a better product than the patent machines.

One of the major problems with brickwork was dampness of the walls, unless the bricks were burnt very hard, in which case their colour became disagreeably dark. The problem was solved in England in the

**Figure 16** The Independent Church, corner of Collins and Russell Streets, Melbourne (architects: Reed and Barnes), built by John Young in 1866–67 with bricks in three different colours.
(Photograph by Henry J. Cowan)

mid-nineteenth century by building cavity walls, that is, walls consisting of two leaves tied together with iron ties, so that water that penetrated the outer leaf was drained in the cavity. It is not known exactly when they were introduced, or who invented them. They were first used in Australia about 1870, and soon found universal acceptance.

Damp-proof courses made their appearance, or rather their reappearance, at about the same time. The ancient Romans had used layers of slate in important buildings. This custom was revived in England, but asphalt or lead were eventually substituted for the slate, as being more effective. The same practice was adopted in Australia. Lead became the common material for damp-proof courses around the turn of the century, and it was gradually replaced by the cheaper aluminium and plastic strips from the 1950s on.

## Natural stone for the building fabric

Brick is a manufactured material, and its durability can be controlled, at least approximately. Stone, like timber, is a natural material, whose durability can only be predicted from experience or from exposure tests. In the first century BC Vitruvius recommended:

> Let the stone be taken from the quarry two years before building is to begin. Then let it lie exposed in an open place. Such stone as has been damaged by two years of exposure should be used in the foundations. The rest, which remains unhurt, has passed the test of nature, and it will endure in those parts of the building which are above ground. (Vitruvius 1960, p. 50)

This is still good advice today, if the stone comes from an unknown source, and if one can afford to wait that long. In Britain the reliable building stones came from quarries that had been in use for centuries, but in Australia a record of experience was initially lacking.

One of the first stone structures to be built in Australia in an exposed location, by Francis Greenway in 1817, was the Macquarie Lighthouse at the Sydney Harbour Entrance, using sandstone quarried nearby from the surface of the exposed rock. While the lighthouse was much admired as an architectural feature, the stone deteriorated, and in 1883 it was replaced by an almost identical lighthouse, designed by James Barnet, built immediately behind it (see also p. 120). A photograph of both lighthouses was taken before the demolition of Greenway's structure (Figure 17). There were quite a number of failures due to inadequate durability of the stone in the early years.

The most commonly used Australian building stone in the nineteenth century was sandstone from the Triassic period, which is a good freestone. Most stone buildings in Sydney and Hobart are built from it, and Sydney and Hobart sandstone has been used in Melbourne and in Adelaide. Triassic sandstone is also found near Brisbane; the Queensland Parliament Building was constructed from it in the 1860s (Spry 1983, plate 7).

**Figure 17** Photograph taken in 1883 of Greenway's Macquarie Lighthouse before its demolition, and of Barnet's lighthouse, built to replace it.

(Photograph by courtesy of the late Arthur Baldwinson)

One of the early sandstone buildings was the Sydney Commissariat Store (see p. 25), which was demolished only in 1938. The oldest surviving stone building is Cadman's Cottage, a two-storey house near Sydney's Circular Quay, built for the Government Coxswain about 1816. It is now owned by the National Parks and Wildlife Service, but there are a number of stone cottages in Sydney and Hobart, built probably not much later, that are still inhabited.

The first 'architectural' stone buildings were erected in the 1830s. They include four court houses in Darlinghurst in Sydney (Herman 1970, fig. 105), in Parramatta (Brown 1995, fig. 31), and in the country towns of Hartley (Herman 1970, fig. 107) and Berrima, all designed by Mortimer Lewis. Freeland (1968, p. 95) describes the Darlinghurst Court House as the 'pinnacle of Greek Revival architecture in Australia', and 'scrupulously correct'. It was one of the first Australian buildings with elaborately carved masonry.

Lewis had worked as a surveyor for the British Inspector-General for Fortifications when he applied in 1830 for the job of assistant to the

New South Wales Surveyor-General, John Oxley. In 1835 he was promoted to Colonial Architect, but he had no academic architectural training; and he seems to have obtained the information on the Greek orders from books.

Elizabeth Bay House (Baker 1969, which includes numerous illustrations) is a Regency-style building begun in 1835 as a residence for the Colonial Secretary, Alexander McLeay. It was designed by John Verge who, after a successful career as a London builder, at the age of forty-six came as a free immigrant to New South Wales in 1828.

The Regency style had developed in England in late Georgian times, and by 1835 it was already losing favour there. Most Regency buildings were built of brick that had been given a coating of stucco to make them appear like stone buildings (see p. 50). However, Elizabeth Bay House is a stone building that was rendered, and it has a wooden pediment that was made to appear like stone. It is the Australian early nineteenth-century building most closely resembling an English country house, with a magnificent central space surmounted by a glass dome.

In it is a 'geometric' staircase, formed by cantilevered stone treads built into the wall, an ingenious type of structure of which it is the finest example in Australia. The treads form a helix, and support one another by shell action. There are a few other, shorter geometric stairs in Australia, but most staircases were built with timber treads that are supported by the wall at one end and by a wooden support structure at the other.

The foundation of the University of Sydney in 1850 was a daring decision for a small colony at a time when there were still only two post-sixteenth-century British universities, London and Durham. Edmund Blacket resigned his position as Colonial Architect to undertake the construction of the first buildings in 1854, including a Great Hall that is still large enough for the present university. It was a visionary project, because in 1853 there were only three professors and twenty-four matriculated students.

Blacket started as a surveyor with the Stockton-to-Darlington Railway in England, but he had a great interest in architecture, and spent much of his spare time and several holidays measuring and sketching old buildings. He came to Sydney in 1842 at the age of twenty-five, and was almost immediately commissioned by Bishop Broughton to build a church. He was already an experienced Neo-Gothic architect when he designed the Great Hall; this is freely based on Westminster Hall, which had survived the Westminster Palace Fire of 1834, and is now part of the British Houses of Parliament.

The university authorities had intended that the first buildings should be of brick for the sake of economy (Herman 1970, p. 42), but Blacket persuaded them to face them with sandstone on the outside, and to use stone both externally and internally in the Great Hall (Figure 18). Its roof is a hammerbeam truss of hardwood, lined with carved boards of Australian cedar (Figure 19). The workmanship was excellent; but some of the sandstone pinnacles, from a quarry in Pyrmont, had to be replaced after less than a century.

Subsequent sandstone buildings in Sydney and Hobart were variations on the same constructional themes: plain load-bearing wall construction, rendered stone walls, the use of classical or Gothic details, all with an interior structure of wood (see p. 37), until iron came into use as a load-bearing material.

Later in the century, columns, load-bearing or decorative, were sometimes made of granite, trachyte or marble in buildings otherwise of sandstone (see p. 50). One nineteenth-century Sydney building, the Equitable Life Assurance Society Building (now the National Mutual

**Figure 18** The Great Hall of the University of Sydney under construction in 1858. The architect Edmund Blacket is in the foreground.
(Photograph by John Smith, the first Professor of Chemistry and Experimental Physics)

**Figure 19** The interior of the Great Hall of the University of Sydney today. It is almost exactly as Blacket finished it, except for the modern organ. The trusses are of hardwood, lined with carved boards of Australian cedar.
(Photograph by Sydney University Photographic Services)

Building) in George Street has load-bearing walls of trachyte, an igneous rock of intermediate composition quarried near Bowral (NSW) (Balint 1984, p. 152; Fraser 1989, p. 231).

Other stones have been used where they are locally available. There are numerous buildings throughout the Australian countryside that are built from local sandstone, local limestone, or local igneous rock (Baker 1915; Brown 1937).

Cathedrals and university buildings are likely to survive, because they continue to serve the function for which they were designed. The more handsome office and bank buildings also have a good survival rate, although some may be put to a very different use, such as a fast-food restaurant. However, Sydney has lost most of its stone warehouses and wool stores. In Hobart, where there has been less pressure for more intensive land use, the entire terrace of former warehouses in Salamanca Place is intact; but it now serves the tourist trade (Figure 20).

The other capital cities are not so fortunate in their supply of natural stone. Melbourne's bluestone is a basalt, a basic igneous rock. It forms the fabric of St Patrick's Cathedral, one of the city's dominant buildings even after the advent of the high-rises. It was designed by William

**Figure 20** Former warehouses of sandstone in Salamanca Place, Hobart.
(Photograph by Henry J. Cowan)

Wardell, who studied engineering in England, but then worked for Pugin before migrating to Melbourne at the age of thirty-five. He was almost immediately asked to design the Catholic cathedral, and the first stone was laid in 1860. It is an impressive building, but some contemporary critics thought the bluestone made it look 'funereal'. There were insufficient funds to build the three spires, and when these were added in the 1930s Sydney sandstone from a Hawkesbury quarry was used.

There was no objection to the use of bluestone for warehouses, which made them look formidable, and some of these very durable stone buildings survive.

The Victorian Parliament building (Figure 21), the most palatial in Australia, was designed by John Knight, who trained as an engineer in England, and who came to Australia in 1852, and his junior partner, Peter Kerr. It was begun in 1856, a few years after the gold rush, less than twenty years after the foundation of Melbourne. It has a bluestone plinth almost a full storey in height. The walls of the two chambers are bluestone, and the building was initially used without a formal facade, for which bluestone was considered 'too gloomy'. The classical facade

**Figure 21** Victorian Parliament House, designed by John Knight and Peter Kerr, begun in 1856. The dome was not built.

(Illustrated in Garran [1886, p. 544])

was originally built of Darley sandstone, from a quarry just outside Melbourne, but some of that deteriorated, and replacement stone came from a quarry at Stawell, 200 km further north-west. This also gave trouble, and further repairs were made with Sydney sandstone. The dome was not built (see p. 57).

Joseph Reed's Victorian State Library, started in 1853, was constructed with sandstone imported from Tasmania (see also p. 92), as were the Law Courts, the Post Office and several bank head offices.

Adelaide has the only major building in Australia whose fabric is mainly of marble. Marble is a metamorphic limestone—that is, one changed by high temperature and pressure in the earth's crust into a fine-grained material that can be cut precisely. The Greeks had local supplies and used it for important buildings, and the Romans employed it with increasing frequency after Augustus established the Carrara quarry that is still in use today.

The South Australian Parliament building has a base structure and steps of West Island granite, and a superstructure of grey marble from Kapunda, about 80 km north of Adelaide (Morgan & Gilbert 1969, p. 11; Spry 1986, p. 32).

The Town Hall, the General Post Office, and the Law Courts are built of South Australian sandstone, of a geological period different from Sydney sandstone, but some other buildings, such as the Bank of Adelaide, were built with sandstone imported from Sydney.

Fremantle has local limestone deposits that were used for the Fremantle Gaol; but Perth is predominantly a brick city (Oldham & Oldham 1961, p. 76). Even Edmund Blacket, who favoured stone for Gothic churches, designed the Anglican cathedral in brick.

The main local stone in Brisbane is porphyry, a fine-grained igneous rock with some large isolated crystals. However, sandstone from the Helidon quarries near Toowoomba, 100 km to the west, or Sydney sandstone, was used for those parts of buildings that required dressed stone. The Anglican Cathedral of St John is built of porphyry from a quarry 8 km from the building, but Sydney sandstone was used for the curved mouldings and the window tracery (Byrne 1995). It is the only Gothic church in Australia with a stone-vaulted ceiling (see p. 55), using Helidon sandstone.

Norfolk Island was occupied as a penal settlement from 1788 to 1815. When it was abandoned, the buildings were destroyed so as not to encourage any other, possibly hostile, settlement. The penal colony was reopened in 1825, and a new set of buildings were erected, using the local limestone (Cox & Stacey 1971, pp. 22–40). This is porous, and most of the buildings were rendered to make them weatherproof. These simple, well-proportioned buildings remained when the penal settlement was finally closed in 1856, and the island was handed over to some of the Pitcairn Islanders, descendants of the Bounty mutineers and Tahitian women, who had multiplied beyond the capacity of Pitcairn Island to support them; some of the buildings are still in use.

## Decorative use of natural stone

The Egyptians and the Greeks always treated their columns as load-bearing, although they were usually so grossly oversized that the stresses in them would have been small. They also usually cut the columns from the same stone as the rest of the building.

On the post-Augustan era the Romans increasingly used costly stones, often highly polished, for columns. Because their surface might be damaged if the columns were installed during construction, the columns in some buildings lost their load-bearing function, so that they could be slid in when the building was virtually complete to avoid damage to their surface. The architraves or arcades, which they seemingly supported, were then carried by the walls.

In the later nineteenth century this Roman method of using polished stone became popular for Australian prestige buildings, Some of these were given exterior columns of granite, trachyte or marble, and decorative stones were also used for interior columns, for staircases, as a veneer for walls, and as floor tiles.

Australia has a wide range of marbles and other decorative stones (Baker 1915, pp. 12–28; Spry 1986, pp. 32–91), and there are some fine examples of their use in nineteenth-century buildings. However, none of the quarries are located near the big cities, so that imported stones can compete effectively, and only three small Australian marble quarries are in production today (Spry 1986, p. 32).

## Other traditional building materials

External plastering is applied to brickwork, and occasionally to natural stone, to make it appear smoother and/or make it more resistant to rainwater penetration. In the very early days of the colony clay had to be utilised (see p. 22), but as soon as more lime was available, lime mortar (that is, slaked lime and sand) was used.

Limestone and seashells, both calcium carbonate, are raw materials for quicklime (calcium oxide), which is made by burning them in a kiln. Slaked lime (calcium hydroxide), also called lime putty, is produced by adding water.

Since lime mortar dissolves slowly in water, it has to be renewed from time to time when used externally. Hence cement mortar is used today on the outside of buildings. Cement was imported from England by the 1840s (see p. 81), but it was used mainly for foundations and hydraulic works until the 1870s.

When plaster was used decoratively, it was often called stucco. External stucco was particularly popular in Melbourne after the gold rush for turning brick buildings into imitations of elaborately-carved stone buildings. Melbourne had the money to import Italian workmen skilled in stucco work, and it had the local supply of rock-lime, which was still in short supply in Sydney. Natural cement was also used from the 1860s on (see p. 81).

Internal plastering was originally done with lime mortar, often with an admixture of hair or vegetable fibre as reinforcement. From the 1830s plaster of Paris was used for interior decorative work, or it was mixed with lime putty.

Plaster of Paris is usually made from gypsum (hydrated calcium sulphate) by driving off some of the water (which makes it a hemihydrate). This material is also found in nature, and the first deposit that came to the attention of builders was near Paris, hence the name. Plaster of Paris expands on setting, and therefore produces a precise casting, which makes it particularly suitable for complicated decorations.

Gypsum was found in South Australia in the 1850s, soon after its first settlement, but it was not mined until the 1880s (Capon 1991, p. 11), and for half a century gypsum was imported from Britain.

The main use of the plaster decorations was for cornices to provide a neat joint between the wall and the ceiling, and for decorative mouldings, such as ceiling roses around pendant lamps. When gas light was introduced (see p. 115), ceiling roses were often made with perforations to allow fumes to escape into the roof space. Some of the plaster decorations were very elaborate, covering the entire ceiling and sometimes also the walls, and they could be quite heavy, so that their support, usually by fixture to the ceiling joists, required attention.

The more elaborate decorations were often prefabricated. This was facilitated by the importation in the 1850s of reusable gelatine moulds, from which precise imprints could be obtained. Canvas plaster (plaster of Paris reinforced with canvas) was introduced about 1885.

The traditional plaster surfaces were originally supported by a timber structure (lath and plaster). Around the turn of the century, metal lath was increasingly used instead of timber. In Melbourne from about 1910 plaster sheets were reinforced with longer fibres of flax, hemp, or sisal; this made them largely self-supporting. The introduction of fibrous plaster coincided with a change of taste (now called the Federation Style) for less ornate, and therefore less heavy decorations; this was well suited to the new technology.

Fibrous plaster, originally a New Zealand invention, was first used in New South Wales about 1916, and by the early twenties Australia-wide (Freeland 1968, p. 222; Capon 1991, p. 42). Fibrous plastering became a separate trade after a demarcation dispute, in which the carpenters had claimed the right to nail the sheets, and the plasterers to stop the nail holes. Eventually it was agreed that fibrous plasterers should do the entire job, provided they did not use a carpenter's claw-hammer. Fibrous plaster thus became an economical material, which retained its dominance until it was replaced by gypsum drywall construction, using factory-produced plasterboard.

Pressed metal ceilings were introduced in the 1880s by Ernest Wunderlich, an English migrant who imported and manufactured a range of building materials in Sydney. They were initially made from zinc sheet, and later from galvanised steel sheet. The embossing process limited the depth of the projections, so that the decorations were more restrained than those on most plaster ceilings of the same era. The ceilings were subsequently also made in Melbourne, and continued in use for about thirty years; but not many survived, because it was difficult to make alterations involving the ceiling without damage to the metal. By contrast, plaster ceilings could easily be cut and repaired.

One of the first zinc ceilings to be built, that of the Sydney Town Hall Auditorium, an 1888 Centennial extension of the building, is still in good condition. Susan Bures (1987, p. 25) relates that Ernest Wunderlich, himself an accomplished amateur musician, persuaded the City Council with a statement that, as the organ to be installed would be the largest ever built, the unpredictable vibrations from a crescendo might well cause pieces to fall off a plaster ceiling.

The earliest roof covering of thatch or bark was not used for long in urban areas because of the fire risk. Timber shingles became the favoured roofing material (see p. 16), because the local clay tiles were initially not entirely successful in stopping rain penetration.

Welsh slates were imported from the 1830s, many as ships' ballast. The size of slates is traditionally denoted by names ranging from 'Ladies' to 'Queens'. The imports to Australia were mostly Countess slates measuring 20 inches × 10 inches (500 mm × 250 mm). The Welsh slates were thin, so that only a comparatively light roof structure was required, but they were a dark colour. Deposits of slate were soon found in South Australia; however, like the local marbles they had difficulty in competing with the imported material. The same happened to later finds in various parts of New South Wales and Victoria.

Corrugated iron sheets were imported in the late 1830s, and they were soon so widely used that they came to be regarded as a vernacular material. They are discussed further in chapter 3.

The first Marseilles pattern tiles were imported by a French company in 1887, and they were an immediate success. They were well-burnt and durable, they interlocked so that only a small overlap was needed, and they had a system of channels that ensured good drainage even in a heavy downpour. When World War I interrupted shipments, the Wunderlich Company manufactured them locally.

Wooden shingles were combustible, slates were expensive, and iron and steel sheets rusted. Thus the Marseilles tile became the dominant roofing material in the better suburbs, until it was replaced gradually by the cheaper concrete tile (first made in the 1920s) in the years following World War II.

One of the few building materials carried in quantity by the First Fleet was glass (see p. 3), and all glass continued to be imported, mostly from England, Belgium or America, until the Australian Window Glass Company started local manufacture in 1932.

The glass that came with the First Fleet was crown glass (see Cowan 1977, p. 206), which could only be made in relatively small sheets. From the late 1830s most of the imported glass was cylinder glass, made by a French-invented process still dependent on glass blowing, but allowing the production of larger sheets (McGrath & Frost 1961, p. 25). It was introduced into England by Robert Chance, the manufacturer of the glass for the Crystal Palace. The first modern process, which drew the sheet directly from a pool of molten glass, was invented in Belgium in 1904, and this soon became the standard method of manufacture worldwide.

## Vaults and domes

The two main traditional building materials were timber and various masonry materials, such as brick and stone. The simplest way of spanning a gap is to use horizontal pieces of material, but beams and slabs develop tension on their lower face (Cowan 1993, p. 52). Timber is equally strong in tension and compression, and thus good for beams, but it can be destroyed by fire. Masonry materials are fire resistant, but while they are strong in compression, they lack tensile strength.

The Romans solved the problems by adapting arched construction, which had been used for centuries in non-durable mudbrick structures (Cowan 1993, p. 51), to masonry structures. They thus developed durable vaults and domes, in which the material was entirely in compression. As a result they were able to build fire-resistant structures with spans in excess of 40 metres, which were not exceeded until iron came into use as a structural material in the nineteenth century.

The Roman arched construction was the basis of Byzantine and Islamic domes, of Gothic vaults, and of Renaissance domes. Many of these are still considered architectural masterpieces unsurpassed in modern times.

By the end of the eighteenth century fire-resistant construction, which had been so important in Roman and medieval times, was no longer as significant. Lightning conductors had been invented (see p. 35), fire brigades came into existence after the Great Fire of London in 1666 (see p. 16), and firebrands to set combustible roofs alight no longer played the same dominant part in warfare. Incombustible roofs also became vulnerable after the development of artillery.

However, vaults and domes retained their prestige because they had featured in so many beautiful buildings of the past ages, and they were used mainly for that reason in the colonies. They were built in Latin America, in India (the Indo-Saracenic architecture during the British Raj), and in the USA; indeed the dome that appears most frequently in pictures today is that of the Capitol in Washington.

Australia was settled later, and the opportunities for building domes and vaults were therefore fewer before the architectural fashion changed.

There is only one Gothic church in Australia with a complete stone vault, the Anglican Cathedral of St John in Brisbane. It was designed in 1886 by an English architect, John Pearson, who had built many Gothic churches in Britain, and it was then redesigned by his son Frank when the site was changed. It has had several architects since, most recently Conrad & Gargett, and it is still unfinished. The earlier part is in traditional construction, but the more recent part has a steel frame. The entire vault is self-supporting in the Gothic tradition, and it is of Helidon sandstone from Toowoomba.

Most Australian Gothic churches, including the cathedrals, have been built with timber roofs visible from the inside, and in spite of the lack of protection from a stone vault only a few roofs have been destroyed by fire.

There are some vaults built of other materials visually imitating Gothic stone vaults. St Peter's, on Cooks River in New South Wales, was built in 1838 with a plaster vault (Cox & Lucas 1978, p. 175).

The most notable example was the first St Mary's Cathedral in Sydney, whose foundation stone was laid by Governor Macquarie in 1821. Patrick O'Farrell (1971, p. 4) considers that Father Terry made the basic design. The cathedral was dedicated in 1836, but further extensions were contemplated when enough money would be raised.

The cathedral had sandstone walls with stone statues. The columns were of ironbark clad with cedar. There was a groined vault in the traditional Gothic manner, but it was built of polished cedar (O'Farrell 1971, p. 16). Above that was a timber roof structure covered partly with shingles (the older part of the building), and partly with slates. The *New South Wales Magazine* of July 1843 commented:

> The interior of the cathedral surpasses the exterior in beauty; the pillars, 40 feet [12 m] in height, are slender and graceful. The groined ceiling, with its broad moulded and arched ribs, its vaulted compartments and intersections, with their ornamented caps and bosses or corbels, which are all enriched with foliage, figures and shields emblazoned, is executed in a very superior style, and has a very pleasing effect. It will remain as an enduring specimen of the mechanized skill and talent in the colony. (Latta 1986, p. 117)

Imitating the stone vault in timber ran counter to the Gothic principle of providing a fireproof barrier. An open timber roof might have survived a fire at least partially; but the layer of thin wooden boards forming the 'vault' made the total destruction of the roof a certainty if the boards caught fire, as they did on 29 June 1865. The fire started about an hour after an evening service had finished; it seems to have been an accident caused by a lighted candle. Paintings or etchings of the fire, which left only the stone walls standing, have been published by O'Farrell (1971, p. 101), Adrian (1984, p. 22, and Latta (1986, p. 119).

Australia's largest dome was also destroyed by fire. There is a difference between using timber for a Gothic groined vault and for a dome. The groined vault is simply a fire barrier under a timber roof, which sheds the rainwater. In Classical and Renaissance architecture the dome was the actual roof, and there were precedents for building it of timber, which meant, of course, that the roof was not fire resistant. However, some survived for many centuries; the Dome of the Rock in Jerusalem, built in the eleventh century as a replacement of a seventh-century dome, and still in existence, is an example (Natshen 1980).

The 'Garden Palace' was built for the Sydney International Exhibition, which was open for 185 days in 1879 and 1880. It was visited by over one million people, at a time when the entire population of Australia was just over two million (Latta 1886, pp. 136–44; Wade 1979). The building was designed by the Colonial Architect, James Barnet, and built by John Young in eight months (see also pp. 83 and 119). It was cross-shaped, and the crossing of the nave and the transept was surmounted by a 100 foot (30 m) diameter dome (Figure 22). This was formed by 32 ribs, each a triangulated truss of Douglas fir, imported from America, and covered with timber planks and corrugated-iron sheets.

It had been intended that this 'temporary' building should be demolished after the closure of the exhibition, but it had become so popular that it was decided to retain it for other exhibitions. However, the huge space in the building was also used by various government departments, and for the storage of public records. The building was a fire hazard because of its softwood structure and the large quantity of paper stored in it. It caught fire in the early morning of 22 September 1882, and the cinders set alight the roofs of several other buildings, one in Potts Point, 1½ km distant. Nobody was killed, but some irreplaceable records were lost in the fire.

The Melbourne International Exhibition opened in 1880 in a stuccoed brick building surmounted by a timber dome, designed by a private firm of architects, Reed and Barnes, and built by David Mitchell

**Figure 22** Contemporary drawing of the 'Garden Palace', Sydney, 1879. The dome had a diameter of 30 metres and a height of 64 metres; for comparison the dome of St Paul's Cathedral in London has a diameter of 34 metres and a height of 65 metres
(Fletcher 1987, p. 1032).

(see also p. 120). The dome has a diameter of 60 feet (18 m), and is also built of imported American Douglas fir. The building is still used for exhibitions and conferences, and it was the venue for the opening of the first Australian parliament.

There are quite a number of smaller Australian domes built for decoration in the later nineteenth century with timber frames: for example, the Bathurst Court House (Figure 23).

A number of large domes have been projected, but never built. The Victorian Parliament House (Figure 21) and the South Australian Parliament building were designed to have domes. Figure 21 was originally published by Andrew Garran (1886, p. 544), and R. T. Baker (1915, p. 81) published a picture of the South Australian Parliament (see also p. 49) with a dome. Both evidently assumed that it was only a matter of time before the domes were built.

In 1861 the New South Wales Government held a competition for a new parliament house, and several of the schemes, although not the one that received the first prize, had large domes. However, due to lack of

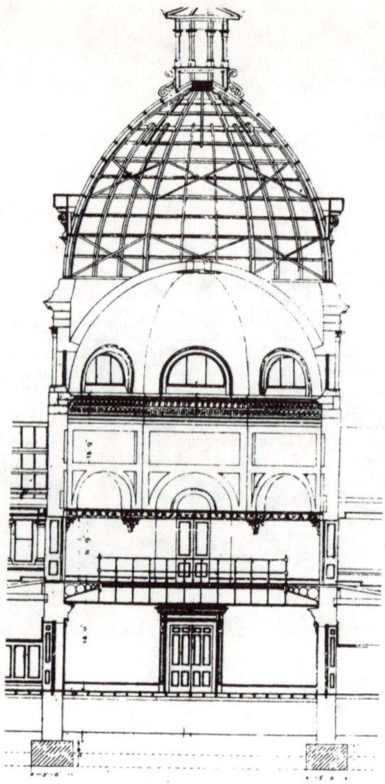

**Figure 23** Cross-section through the dome of the Court House in Bathurst, a town 200 km from Sydney. It was designed by the Colonial Architect, James Barnet, and constructed in three stages between 1877 and 1883. The span of the timber dome is 30 feet (9.1 m).

(Reproduced from a drawing in the archives of the Bathurst Historical Society, by courtesy of the Society)

funds the 1811 Rum Hospital building survived (see p. 28). The same happened in 1896 after Walter Vernon, the Government Architect, proposed a new building surmounted by an even bigger dome (reproduced by Pegrum 1983, p. 18). The additional office accommodation needed for the Parliament was eventually provided in the 1970s by an extension at the rear of the existing building that left the old facade intact.

Australia thus never built a dome of international ranking in traditional permanent construction. However, in 1911 it set a world record for the span of a reinforced concrete dome (see p. 94).

# 3

# IRON AND STEEL

## The Industrial Revolution and Australian construction

Wrought iron was first produced about 1400 BC. As the technology to produce a temperature high enough to melt pure iron did not exist until the nineteenth century, the ore was heated by charcoal to reduce it to a soft iron sponge. This almost pure iron was then worked (or wrought) into the desired shape at white heat.

The other traditional form of iron is cast iron, which contains 2–4 per cent of carbon, absorbed from the charcoal. This lowers the melting point by about 250°C, and this temperature could be produced in a charcoal fire with bellows. In China it was first produced in the sixth century BC, but in Europe not until the twelfth century AD.

Iron remained an expensive material until 1709 when Abraham Darby in the west of England smelted iron ore with coke, which made it possible to produce much larger quantities of iron, and at a lower cost.

The first structure completely of iron was a bridge, built in 1779 by Abraham Darby III near his ironworks in Coalbrook Dale, to show how light a structure could be built with the material (Cowan 1977, p. 246).

Iron was the prerequisite for the development of engines using high-pressure steam. In 1769 James Watt patented the condensing steam engine, and the first of these engines were usefully employed in the 1770s. Machines for producing power, such as windmills and waterwheels, had been used for centuries, but they had limitations imposed by these natural resources. With steam power it became now possible to supply as much power as might be needed, and where it was needed.

The new factories were generally built by engineers whose main expertise lay in the installation of the all-important engines that provided the power, and who were therefore experienced in the use of iron. Iron structures were developed because protection against fire was a major concern (see p. 35). They combined the advantages of the traditional timber and masonry structures, good tensile strength *and* incombustibility (see p. 54). The first use of iron beams in a factory building was in Belper (Derbyshire) in 1811 (Cowan 1977, p. 250).

These technical innovations of the Industrial Revolution occurred at approximately the same time as the beginning of transportation to New South Wales, some before and some afterwards. They soon began to affect life in Australia. The first steam engine was brought there in 1813 by a free settler, and it provided the power for a flour mill. A steam-powered timber saw was used in 1838 in New South Wales, and in 1841 in the Port Phillip District (Melbourne), which had only been settled in 1834. In 1835 the first boiler was fabricated locally, and in the 1870s hydraulic power, transmittable through pipes, came into use in Australia (see p. 175).

The First Fleet had carried ten tons of iron sections, as well as nails and spikes (see p. 3). More iron came on subsequent ships, and a local iron industry started in the 1840s.

## Wrought iron, corrugated iron, and galvanising

In the factories that produced cast iron the liquid iron was generally run into a sandbed, with small channels running off the main channel; this was thought to be reminiscent of a sow suckling her piglets. After cooling, the metal was broken up, and the 'pigs' became the raw material for the precision castings required, and also for a new and more economical process for making wrought iron.

In 1784 Henry Cort patented in England the less laborious 'puddling process' for making wrought iron, which reduced the carbon content of pig iron by heating it in a 'reverberatory furnace', a shallow tank in which the iron was kept separate from the fuel so that it could not absorb any carbon, and the heat came from flames drawn over the tank.

Fabrication also became easier. In medieval times wrought iron had to be formed by hammering. In the eighteenth century sheets could be produced by passing the red-hot material through rollers, operated at first by a waterwheel, and later by a steam engine. Using this new source of power, angle sections were first rolled in 1820, T-sections in 1830, and I-beams in 1848. The development of the now familiar I-beam was an important innovation, because until then iron girders had to be assembled by riveting from iron plates and angles (see p. 69 and Figure 30). Riveting, which had been done by hammering in medieval times, also became cheaper and faster through the use of mechanical power.

While the structural use of wrought iron did not affect Australian construction until the later part of the nineteenth century (see p. 69), another innovation had more immediate influence.

With steam-powered rollers it became possible to produce much thinner iron sheet. This material was impermeable to water and it did not burn, but it was too flexible for use in a wall or in a roof. In 1829 Henry Palmer, a London civil engineer, patented 'the application of fluted, indented or corrugated metallic sheets or plates to the roofs and other parts of buildings' (Herbert 1978, p. 34), and sold the patent to Richard Walker, a manufacturer of prefabricated buildings (see p. 65).

In the late 1830s corrugated-iron sheets were exported to Australia, where they became so popular that it is today often forgotten that they were not a native material, and were not produced locally until the twentieth century. Part of the widespread early use of corrugated iron was due to the *New South Wales Building Act* of 1837, which had forbidden the use of thatched and shingled roofs for reasons of fire safety. When corrugated-iron sheets became available, they were quickly adopted because they were the cheapest permissible form of roof covering.

Rusting is a major problem in the use of iron, particularly when it is in the form of thin roof sheets. The oxidation of metals generates heat, so that the process is self-perpetuating, unless the supply of oxygen is cut off. This happens when lead or copper is used for roof sheets. Lead absorbs carbon dioxide from the atmosphere to form a hard layer of lead carbonate, and copper forms a hard layer of copper oxide; both stop further oxidation. Iron oxide, however, combines with water vapour to form rust (iron hydroxide), which is soft and water soluble, so that the oxidation continues until all the iron has turned into rust.

In the Gothic cathedrals iron bars were used to support and stabilise the great windows that consisted of pieces of glass joined with lead.

The various processes to prevent rusting were shrouded in secrecy, but one of them evidently consisted of coating the iron with zinc. This was within the competence of medieval craftsmen, because the melting point of zinc is only 420°C.

In 1837 French chemist Ernest Sorel rediscovered the process, and adapted it to mass production (Elliott 1992, p. 82). Sorel coined the term galvanised iron, because he considered that the bond between iron and zinc was due to galvanic action.

Galvanising greatly improved the usefulness of corrugated iron roof sheeting, and by the early 1840s most of the sheets imported into Australia were galvanised. Galvanised iron also led to the creation of a new industry for the manufacture of prefabricated iron buildings (see p. 64).

## Cast iron for light construction and for decoration

While corrugated iron sheets were imported during the whole of the nineteenth century, cast iron was produced locally from the 1830s on. However, the pig iron came originally from England, and even after a local iron industry was started in 1848 (see p. 67), the majority of the pig iron was still imported.

In the 1830s there were already several well-equipped foundries in Sydney. Ian Jack and Aedeen Cremin (1994, p. 15) list some of the equipment of William Browne's foundry in Sussex Street: a 5-horsepower steam engine, a furnace assisted by a pair of large bellows, an overhead crane to transport the ladles, mould boxes and castings, a drilling machine, a boring mill, and a lathe with a 6-metre bed.

Melbourne's first foundry was established by Robert Langlands and Robert Fulton in 1842 (Carroll 1988, p. 51). It was then still a small town, but it grew rapidly after the gold rush in the 1850s, and cast iron was used extensively in its new houses (Turner 1985, p. 64).

One of the initial incentives for the use of cast iron was its incombustibility. It was thus employed for stoves and fireplace grates, and also for the columns of multi-storey buildings used by the public, such as theatres, shops and churches. Adrian (1984, p. 30) points out that theatres were 'the most fire-prone establishments in Sydney's early days'. Thorne (1995) cites fires that caused the destruction of the Theatre Royal, the Royal Victoria Theatre and the Prince of Wales Theatre in Sydney, and of the Theatre Royal, the Haymarket Theatre, and the Bijou Theatre in Melbourne, all between 1840 and 1889 (see also p. 16).

However, the decorative nature of cast-iron columns, balustrades and spandrels soon became the major reason for their use (Figure 24). Decorative gates and screens had been made in wrought iron at least since medieval times; but the labour content was very high, as each screen had to be hot-worked individually. In cast iron the same pattern could be repeated many times using a single prototype.

The first decorative cast iron was imported from England. The columns were generally modelled on the classical Orders, particularly the Corinthian, and the panels were decorated with a variety of geometric and floral patterns. It was soon found economical to produce the castings in Australia using, however, mostly imported pig iron (Figure 25).

Decorative cast iron was used particularly in Victoria, New South Wales, and Queensland (see also Figures 41 and 42); its popularity declined after the turn of the century. The cast iron of Melbourne is especially well known because Dr Graeme Robertson, a neurologist at the Royal Melbourne Hospital, and his daughter Joan have documented its production by more than forty-two foundries, and published many excellent photographs (Robertson 1967; Robertson & Robertson 1975; and Robertson & Robertson 1977).

**Figure 24** Boronia House, Mosman, NSW, a residence built in 1885 for John and James Kerry, now a restaurant. Single houses were often built with cast-iron verandas surrounding the house. However, most of the cast iron was used in terrace houses, in the front of the building only.

(Photograph by Henry J. Cowan)

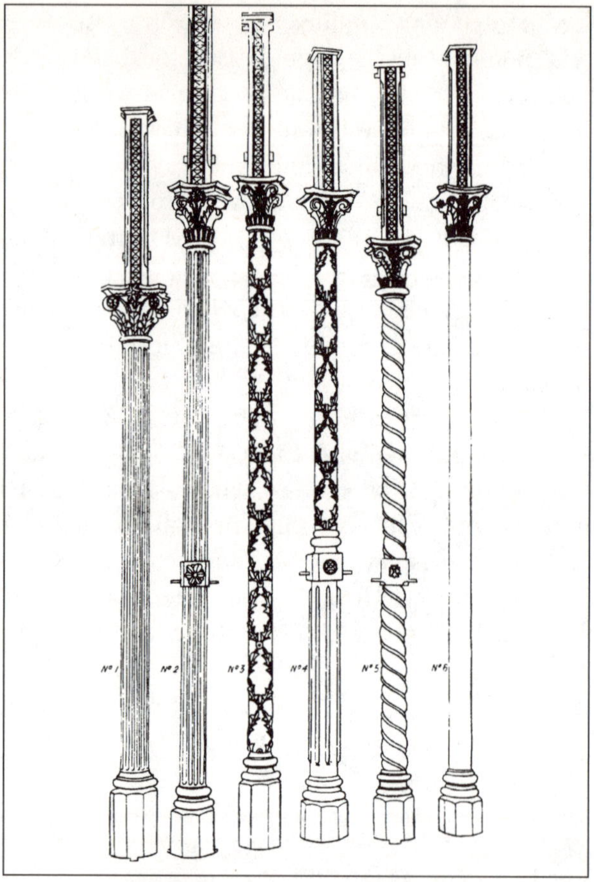


**Figure 25** Page from the catalogue published by the Excelsior Foundry in South Melbourne in January 1901.
(Reproduced from Robertson [1967, p. 55])

## Prefabricated iron houses

The first 'Exhibition of the Industry of all the Civilised Nations of the World', held in the Crystal Palace in 1851, included a number of prefabricated iron buildings. The exhibition building itself was an iron building that had been partly prefabricated, and demonstrated a possible new approach to building construction (see also p. 160).

In Victoria after the gold rush of 1851, which caused a great increase in population, and also in South Australia, which had only been settled in 1836, people wanted houses more quickly than they could be built locally. They had the money to pay for them, and in Britain there were manufacturers willing to meet the demand.

A large number of prefabricated iron houses was therefore imported during the 1850s, particularly into Victoria and South Australia. The development of corrugated-iron sheeting since the mid-1830s (see p. 61) had made it possible to provide all-iron houses that were relatively light, easily assembled, and in addition fire and termite resistant. Cast iron was used for buildings of more substantial construction.

About a dozen firms manufactured metal houses for export (Herbert 1978, pp. 40–74). Some specialised in the field, some also produced prefabricated wooden houses (see p. 37), and some also manufactured other iron products. The principal manufacturers of prefabricated iron houses were Richard Walker, of London, who regularly advertised in Australian newspapers; Edward T. Bellhouse, of Manchester; and Samuel Hemming, of Bristol, whose son emigrated to Australia and represented him there.

In addition, some prefabricated iron houses were made locally, notably by Peter Johns in Flinders Lane, Melbourne (Blainey 1956, p. 8).

Australia was just one, and not the most important market for British prefabs. There was a steady export market to South Africa and to various parts of Central and South America where British companies were active, and a large number had been exported to California during the gold rush of 1849.

Some of the residential buildings were quite luxurious (Figure 26), and so were some of the public buildings. Prefabricated buildings were ordered for a number of shops, hotels, restaurants, and theatres. The Olympic Theatre, built to order for George Coppin by Bellhouse, was erected at the junction of Lonsdale and Stephen (the present Exhibition) Streets in Melbourne, and it became well known.

In Sydney the NSW Legislative Council in 1857 was housed in an iron prefab imported from Melbourne. It was constructed as an extension of the Surgeons' Building of the Rum Hospital (see p. 28), when the NSW Legislative Assembly was established, and took over the Council's previous meeting room.

Several prefabricated iron churches were also built, and some survived into the twentieth century. The majority were in Melbourne, but one of the most ornate was the 'Iron Church' in Macquarie Street, Sydney, which had a two-storeyed Ionian colonnade and two towers; it was manufactured by Charles D. Young in Glasgow, and erected in 1855 (illustrated by Herbert 1978, p. 105).

**Figure 26** House designed by the architect James Henderson, and made by the iron founders Goldie & Inglis for erection in Melbourne in 1853. It had cast-iron columns and window frames, a corrugated-iron outside lining for the walls, tongued-and-grooved deal inner linings and partitions, and mahogany window frames and sashes. The use of curved corrugated sheeting, which gave extra stiffness to the roof, was not uncommon.

(*The Civil Engineer and Architect's Journal*, 1853, p. 456; and reproduced from Herbert [1978])

The age of iron prefabs lasted less than twenty years, and nearly all the buildings have now vanished. One survives at 388 Coventry Street, South Melbourne, manufactured by Walker and erected in 1853; it has an exterior entirely of corrugated iron sheets (illustrated by Ogg 1987, pp. 122–3). Another is Corio Villa, erected in Geelong (Victoria) in 1854, a cast-iron building made by C. D. Young (illustrated in Saunders 1966, p. 172).

## Australian and imported iron and steel

During the nineteenth century most of the iron and steel used in Australia was imported, but there were several attempts to produce iron from local ores.

The raw materials for the production of 'pig iron' (see p. 60) are iron ore, coal, and limestone to serve as a flux. By 1806 all three materials had been found. Indeed, Australia has since become one of the world's largest exporters of both coal and iron ore.

The first iron-smelting operation commenced in 1849. The capital for the Fitzroy Iron Works near Mittagong (NSW) was provided by a local landowner and three Sydney businessmen, and a small furnace was built with guidance from an English ironmaster. The plant was quite successful considering its small size, but the cost of producing the iron was greater than that of iron imported from Britain, which came mostly as ballast in ships serving the growing wheat trade. The Mittagong plant operated until 1852. It was then closed, and eventually re-equipped, and operated from 1864 to 1872, and a rolling mill was built. It had two more short periods of operation, and the last smelting near Mittagong was done in 1891.

The other early iron-smelting operations had an even shorter life. Three smelting works were established in the Tamar Valley near the present-day Launceston (Tasmania), another at Lal Lal near Ballarat (Victoria), and further blast furnaces were built at Mt Jagged on the Hindmarsh River south of Adelaide, and at Bogolong in New South Wales. All started and were closed down in the 1870s. These operations have been fully documented by Jack & Cremin (1994, pp. 14–95).

In the 1870s one of the main uses of iron was for rails. The structure of multi-storey buildings was mostly of timber (see p. 36), and road bridges were also mostly built of timber. However, when the building of the Australian railways and tramways started in 1854, there was a greatly increased demand for iron. Rails for Sydney's trams (see p. 119) were the main output of the first successful iron works, which was established at Lithgow (NSW) in 1876 by Enoch Hughes and James Rutherford.

In 1886 William Sandford, the new owner of the Lithgow Ironworks, installed the first Australian steelmaking plant, a 4-ton Siemens Open Hearth Furnace. Whereas wrought iron is almost pure iron, steel contains 0.1–1.0 per cent of carbon, which greatly increases its strength. It had been a very expensive material reserved for weapons and tools, and forged laboriously from *wrought iron*, until in 1856 Henry Bessemer invented a 'converter' that blew hot air through *pig iron* to 'burn out' its undesirable constituents. However, by the time Sandford installed his plant, this had been superseded by the Siemens process. The Lithgow plant also ran into financial difficulties, but it survived (Figure 27), and in 1922 it was moved to Port Kembla on the coast.

In the meantime the Broken Hill Proprietary Company (BHP) had decided to establish a steelmaking plant in Newcastle, which had large coal resources. Since 1885 BHP had mined from the Barrier Ranges

**Figure 27** Ruins of the Lithgow steel-making plant. The building in the foreground was erected in 1906 to house a Davy engine, which supplied the power.
(Photograph by Henry J. Cowan)

silver–lead–zinc deposits whose life was limited (McLennan 1974, pp. 3–16), it owned high-grade iron-ore deposits, and it had the capital to establish a large modern plant whose products could compete with imports.

The Newcastle Steel Works opened in 1913. World War I, which broke out the following year, cut off most imports, and created a demand for locally produced steel. By 1918 the plant's steelmaking capacity had been doubled. By the 1930s it was claimed that 'Australia produced the cheapest steel in the World' (McLennan 1974, p. 35), whereas most other manufactured products were more expensive in Australia than elsewhere. This was due to the ready availability of the raw materials in easily accessible deposits near deep-water ports, to the economy of mass production, and to the assured market for steel in the construction industry that emerged in the 1920s (see p. 77).

## Iron and steel in multi-storey buildings

The use of hardwood storey posts and beams was almost universal in Australian multi-storey buildings (see p. 36 and Figure 13) until the 1860s.

Even then it preponderated until the 1880s. There was not the same inducement to substitute cast iron for timber as there was for veranda posts and balcony supports, because 12 inch (300 mm) square ironbark columns were much more fire resistant than slim veranda posts. There was also less reason for changing to iron than there had been in England, because hardwoods such as ironbark and jarrah were more fire resistant than any European timber. The use of cast iron for multi-storey, heavily-loaded buildings therefore came to Australia relatively late.

Round cast-iron columns were often made hollow to give them greater lateral resistance, and were then sometimes utilised as down pipes for rainwater. Often they had simple classical decorations. The main alternative was a cruciform section, which was used particularly for heavily loaded columns.

The girders supported by cast-iron columns might be timber or wrought iron, or less commonly cast iron. By 1880 it was generally known that cast iron was not as good a material for girders as for columns, because it was brittle, and consequently had a lower strength in tension than in compression.

The cast-iron members were often cast in Australia, but mostly from imported pig iron. They continued to be made until World War I, but became less common after the turn of the century.

Wrought-iron girders came into use in the 1850s (Balint 1984, p. 17), at first with cast-iron columns, but later the columns were also made of wrought iron. They were often fabricated locally, but some were imported fully fabricated. Large wrought-iron columns and girders were generally assembled from plates and angle or T-sections (Figure 30), because at the time there were no rolling mills in Britain with sufficient power to roll large I-sections.

With the increased use of wrought iron came concern for fire protection of the metal, particularly after a number of major fires in multi-storey buildings in the USA. Iron does not burn in a fire, nor is it likely to melt because few fires reach a sufficiently high temperature. However, wrought iron and steel lose about half their strength at a temperature of about 500° C, which is reached in any serious fire. Cast iron is a more rigid material, and hardwoods such as ironbark are protected for some time by the formation of an insulating layer of charcoal (see p. 35).

While fire protection for wrought iron was not a legal requirement before 1912, architects and engineers increasingly provided it. One method was to cover the iron with terra cotta; this was often moulded with sawdust that burned during firing, leaving a light terra cotta that could be sawn, and was therefore called terra cotta lumber. The other method was to encase it in concrete, which was still a new material (see chapter 4). Cast-iron columns were also occasionally encased in concrete. In addition, the timber floor was often given special protection. The brick jack arches used in Britain (see p. 36.) never became popular in Australia, but in the 1880s an arched structure based on the same principle came into use (Figure 28). It was eventually superseded by reinforced concrete floors (see p. 100).

The construction of the railways created a great demand for rails. These were at first made from wrought iron, but from the 1860s steel rails were rolled in Europe and in America. Rolled steel sections were therefore well established when in 1884 both Dorman Long in England and Carnegie-Phipps in the USA commenced rolling structural steel sections. They co-existed for a few years with wrought-iron sections, but eventually replaced them.

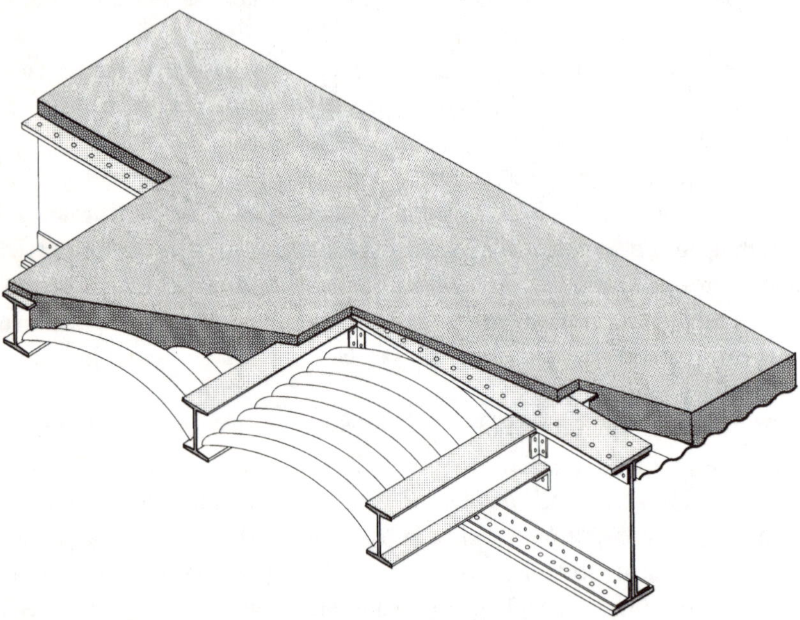

**Figure 28** Fire-resistant floor structure spanning between wrought iron or steel joists, formed by arched sheets of heavy-gauge, deeply corrugated iron and a topping of coke-breeze concrete; these sheets were initially imported from Germany. The softwood timber floor was laid on top of it, and a timber or plaster ceiling was fixed below.

(Drawing by Peter R. Smith)

Although there was some local production, most iron and steel was imported before World War I. Britain had a commanding lead in iron and steel production for most of the nineteenth century, but in the early twentieth century steel was imported also from the USA, France, Belgium and Germany. From 1916 Australia became an exporter of steel; but steel was still being imported, particularly for special structures, such as the Sydney Harbour Bridge.

## The Queen Victoria Building

The transition from cast iron and wrought iron to steel was a gradual process that lasted from the late 1880s to the end of World War I. The Queen Victoria Building, adjacent to Sydney's Town Hall, is of particular interest, because it was constructed with load-bearing masonry walls and columns, cast-iron columns, wrought iron for the smaller structural members, *and* steel in various forms. It was extensively restored in the 1980s, and as a result we are well informed about the structure previously mostly hidden by fire protection and decorative finishes.

The site of the building was set aside as a market by Governor Macquarie in 1810, and it was handed over to the Municipality of Sydney when it was incorporated in 1842. In 1891 the Council instructed the City Architect, George McRae, to draw up plans for a three-storey Queen Victoria Market Building with a deep basement.

McRae was born in Edinburgh in 1858. He was appointed assistant to the City Architect shortly after arrival in Australia in 1884, and in 1887 he succeeded him. He became Government Architect in 1912, and on his retirement in 1923 a collection was made to establish an annual prize at the University of Sydney, which is still the senior award for architectural construction.

The *Australasian Builder and Contractors' News* reported in 1893 (Booker 1979, p. 121) that McRae submitted four designs in the Gothic, Romanesque, Renaissance, and Queen Anne Style, and that the City Council had selected the Romanesque design (Figure 29).

The *Sydney Morning Herald* (1898) commented that 'the design is a modified form of modern Romanesque in which the massiveness, which gives the American Romanesque style so much of its effective force, has been subdued'. Construction started in 1893, and the building was opened in 1898.

**Figure 29** Perspective of the Romanesque design of the proposed Queen Victoria Market Building by the City Architect, George McRae, 1892.

The building was designed to have good daylight. The central avenue had a glass roof supported by semi-circular triangulated trusses. There was a copper-covered, steel-framed dome with a lantern over the crossing with the transept; the *Sydney Morning Herald* (1898) claimed that its diameter of 62 feet (19 m) was at the time the largest in the Southern Hemisphere. Under this outer dome there was an inner dome of stained glass framed in steel. The basement was originally lit during the day by a large number of prismatic glass pavement lights.

The outer load-bearing walls were of Sydney sandstone quarried at Waverley, and in addition there were load-bearing piers of trachyte (an intermediate igneous rock) from Bowral. Trachyte was also used for the cantilevered 'geometric' stairs (see p. 45).

The external walls were carried by 'very heavy' steel box girders over the shop fronts, and these were supported by large trachyte piers.

The *Sydney Morning Herald* reported that

> in order to prevent the possibility of dampness, the whole of the basement is enclosed by damp-proof retaining walls of two thicknesses of 14 in. [0.36 m] brickwork in cement, separated by a vertical damp-proof course at the floor level . . .
>
> The ground floor is constructed with main box girders running transversely, 2 ft 6 in. deep and 1 ft 6 in. wide [0.76 × 0.46 m], over the flanges, and spaced at about 16 ft [5 m] centres. These are supported by two rows of cast-iron columns of 20 in. [0.51 m] diameter and 2 in. [50 mm] metal

(Figure 30), having bracketed spreading caps to which the girders are bolted, the column bases being bolted to large trachyte foundation beds. The street ends of these girders are bolted to specially designed cast-iron stanchions, which are bolted at the bases to trachyte beds.

The floors of the avenue and entrances are immediately carried by 10 in. × 5 in. [254 × 127 mm] rolled steel joists weighing 30 lb per ft [10.44 kN/m] spaced at from 4 ft 6 in. to 5 ft 6 in. [1.4–1.7 m] centres running longitudinally, seated upon wrought-iron chairs riveted to the webs of the transverse box girders; these joists are additionally strengthened and braced by flat iron tie rods passing through and bolted to webs.

The flooring is formed by fire-proof terra cotta lumber 15 in. [0.4 m] deep, carried by and completely encasing the rolled steel joists, the underside being plastered, and forming the ceiling of the basement. On top of this terra cotta lumber, concrete is laid varying from 9 in. to 12 in. [0.2–0.3 m] in thickness, floated with cement, and finished with an elaborate pavement of high-class encaustic tiles. (*SMH* 1898)

The cast iron was cast locally from imported Scottish pig iron and local scrap. The steel and wrought iron were fabricated locally from imported plates and sections. The riveted box girders were used because rolled sections of sufficient size were not yet obtainable at the

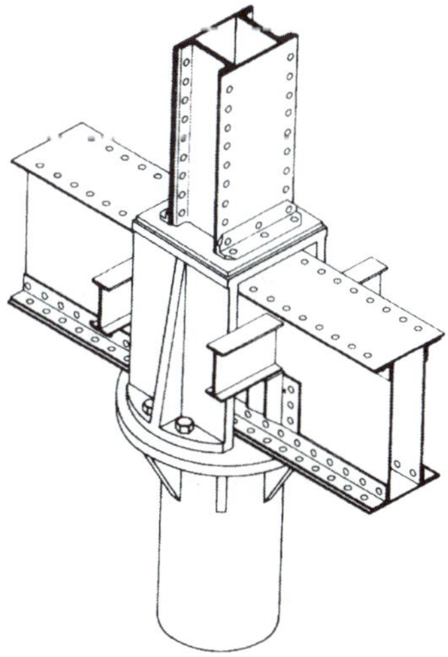

**Figure 30** Detail of the iron structure of the Queen Victoria Market Building, 1898, showing the support of the riveted plate box girders and a riveted plate column on a hollow cast-iron column. (Reproduced by courtesy of the author from Ogg [1987])

time. Rolled steel sections of a much smaller size, however, were used for the joists.

The terra cotta lumber (see p. 70) was fully imported. For the floors specially formed terra cotta blocks were used, which formed a flat arch between the supporting steel joists. All of the stanchions (including those of cast iron), columns, girders and joists were encased in terra cotta lumber. The standard of structural fireproofing was exceptionally good for its time. In addition the building was provided with sixty-four fire hydrants and hoses. The consulting structural and mechanical engineer was George Massey.

The artificial lighting was by incandescent gas lights and gas arc lights (*SMH* 1898)

Hydraulic lifts were supplied and installed by the Melbourne-based Waygood Elevator Company. The goods lifts had a speed of 25 feet per minute (0.12 m/s), and the passenger lifts 150 feet per minute (0.76 m/s). 'Electrical communication of the latest description was provided' (*SMH* 1898). The lifts could have been connected to the mains of the Sydney Hydraulic Power Company (see p.177), but it was decided to install three pumps in the basement, operated by a British Crossley Otto-cycle gas engine, and a hydraulic accumulator packed with 100 tons (102 metric tonnes) of sandstone (see p. 179).

When the building was restored, David Earle, the conservation consultant to the Sydney City Council, said that 'above all else, the Queen Victoria Building is an enormous tribute to the workmen who carried out all the original work'.

## The skeleton frame, and the abortive attempt to build tall

There is a widespread misconception that the skeleton frame was the essential prerequisite for skyscrapers. There is, in fact, a solitary mid nineteenth-century example of a skeleton frame, a boat store in the Sheerness Naval Dockyard in the Thames Estuary, designed in 1858 by Godfrey Green, a military engineer (Collins 1983, p. 56). It is a four-storey building with corrugated-iron cladding, and it probably seemed natural to use a frame of wrought iron and cast iron to support it. It had no influence on subsequent skeleton construction, and it would have been forgotten if a demolition contractor in the 1950s had not noticed that it was a very early skeleton frame. It was not demolished, and is now a Grade I Listed Building.

The high-rise construction, which started in Chicago after the Great Fire of 1871, initially used outer load-bearing walls and inner columns of iron. Load-bearing inner walls could have been used, but they would have taken up more space.

The use of columns to support the load along the outside perimeter, and thus replace the load-bearing walls with curtain walls, was introduced by William Jenney in the Home Insurance Building, a nine-storey Chicago building erected in 1885. Jenney made the decision not for structural, but for constructional reasons: it enabled him to bring materials into the interior of the building during the construction of the upper floors, which greatly facilitated their handling, and avoided obstruction of the public space outside the building. However, the structural advantages of the skeleton frame were soon appreciated.

At that time American building heights were still modest. The term 'skyscraper' was first used in an editorial in a Chicago newspaper in 1891, which enthused about the city's wonderful new tall buildings, and listed ten 'skyscrapers' that were more than ten storeys high.

Melbourne was not far behind Chicago in the late 1880s. The installation of hydraulic power systems in 1887 in Melbourne and in 1889 in Sydney completely transformed the economics of high-rise construction (see p. 176), and reversed the traditional valuation of building space. As long as people had to climb stairs, the upper floors were the least desirable. With lifts they came to be appreciated as being quieter and more private, and in the City of Sydney the harbour views were an added asset. Businesspeople liked the pedestrian access to the people with whom they had to work, made possible by the higher population density. Moreover, landowners could make very large profits from building high.

However, this was used by some people as an argument against permitting high-rise construction, and there were other objections to tall buildings. The fire brigades in Melbourne and Sydney at the time regarded their ability to fight fires in burning buildings from the outside as essential for fire safety. This could not be done with the existing equipment if buildings were higher than 150 feet (46 m), and it seemed unlikely that equipment could be made that would extend this limit.

Tall buildings cast a shadow, but it does not follow that a height restriction creates more solar access. A few carefully regulated tall buildings may be preferable to having a narrow street with buildings all constructed to a maximum permissible height of, say, 150 feet.

The overpowering effect of modern tall buildings on lower historical buildings of great architectural merit is a more serious problem. This did not arise in Chicago, because the 1871 Fire destroyed all but one of the old buildings. New York undoubtedly lost some of its architectural character through the construction of tall buildings, but gained more. In most European cities the conflict was delayed until after World War II, either because regulations were introduced to restrict the height of modern buildings, or because there was a lack of interest in constructing skyscrapers. Australia, which had erected few buildings of outstanding architectural merit before the 1880s, might have been expected to consider itself as part of the New World.

However, the height limit was imposed in 1912 as a result of objections to the 170 foot (52 m) high Culwalla Chambers at 67 Castlereagh Street, Sydney. It was considered an ugly building, and that was at least in part responsible for this decision.

It had been intended that Culwalla Chambers should occupy a large, approximately square piece of land on the corner of Castlereagh and King Streets, but the owner of a four-storey hotel on the street corner refused to sell. The construction of the now L-shaped thirteen-storey building went ahead, on the assumption that the hotel owner, in due course, would sell.

One contemporary photograph showed the completed building with rough brick walls on the nine storeys projecting above the hotel, which were presumably intended to become interior walls after the demolition of the hotel; a later photograph showed these walls covered with advertisements painted on the brick. The building has since been renovated, and it now has tidy walls with windows overlooking the four-storey corner building (illustrated by Adrian 1984, p. 64); but the original appearance would have created support for the aldermen who favoured a 150 foot restriction, which remained in force for forty-three years (see p. 190).

Culwalla Chambers has load-bearing brick walls with stone dressings on the lower storeys. The interior structure consists of steel columns, beams and joists encased in concrete, and reinforced concrete floors (Balint 1984, p. 146).

Later in the same year, 1912, Melbourne made the maximum permissible building height dependent on the street width, which generally limited buildings to slightly less than 150 feet. In the 1920s most other Australian cities also restricted building heights.

The first Australian steel skeleton frame was not built until after the height restriction was imposed. Max Freeland (1968, p. 249) has cited Nelson House, built in 1910 and still standing at 283 Clarence Street, Sydney, as having the first skeleton frame. It is a nine-storey building with concrete-encased steel columns in the outer walls. However, Balint (1984, p. 158) has pointed out that it also has brick piers in both the front and the rear facades carrying some of the vertical load, and side walls that may be presumed to act as shear walls. The building provided the constructional access that motivated William Jenney in Chicago, but from a structural point of view it is not a true skeleton frame.

Balint (1984, p. 186) considers that the accolade belongs to the Commonwealth Bank, built in 1916 and still standing at 120 Pitt Street, Sydney. It is a ten-storey building with concrete-encased steel columns and beams, and the facade columns are also clad in trachyte stonework. The external infill walls are of brick. It has reinforced concrete floors, using expanded metal as reinforcement. Balint points out that taller buildings in Sydney and Melbourne had been erected with loadbearing walls.

The structural theory used for the design of these buildings was straightforward. The joists and beams were regarded as simply supported, and an empirical allowance was made for bending in the columns. When the City Council approved Nelson House, it did so 'subject to the steelwork being of sufficient strength' without attempting to check the design (Freeland 1968, p. 251). However, reinforced concrete, discussed in the following chapter, presented more problems, and in 1924 the City of Sydney appointed J. Rankin to set up a department to check the strength and safety of structures, and specifically to check the reinforcement of concrete.

# 4

# CONCRETE

## The rediscovery of concrete

Modern concrete is an artificial stone that is not soluble in water. Throughout the Middle Ages and the Renaissance in Europe, and in Australia until about 1850 and occasionally later, a 'concrete' was in common use, particularly for foundations and for the cores of massive walls, which like modern concrete contained stone chips or gravel, but whose binding material was water-soluble (see p. 82). In Sydney during the first few years, clay was often used, but the normal material was lime mortar. A few limes have 'hydraulic' properties—that is, they can be used under water because they are not water soluble—but most lime mortars are washed out by water and rain, and they are therefore durable only in locations where this cannot happen.

In the eighteenth century it was well known that the ancient Romans used hydraulic cements (Cowan & Smith 1988, p. 114) to build very durable structures that survive to this day, some of enormous size. However, while the Roman art of building vaults and domes survived in modified form (see p. 54) the fall of the Roman Empire, the art of making water-resistant concrete did not, and attempts during the Renaissance to revive it failed.

Two incorrect statements, which had been handed down from Roman times and were repeated as recently as 1753 by Bernard de Bélidor, a noted French scientist, in his book *Architecture Hydraulique*, may have been the cause of this failure. One was that the strongest limestone made the strongest mortar. The other was that the purest white limestone made the strongest mortar.

In the eighteenth century a number of new lighthouses were built because of the increase in transatlantic navigation. One of these was on the Eddystone, a rock submerged at high tide outside the port of Plymouth (Figure 31). After three timber lighthouses had been destroyed in less than sixty years, John Smeaton was asked to investigate the construction of a masonry lighthouse in this exposed position.

Smeaton decided this required a hydraulic mortar. He investigated the first statement and proved, using the new testing machines developed in the eighteenth century, that the strength of lime mortar made from strong marble was the same as that made from weak chalk.

**Figure 31** Frontispiece of John Smeaton's book on the building of the Eddystone Lighthouse, in which he described his experiments on lime. The artist probably used his imagination rather freely in portraying the lighthouse in a storm.
(Smeaton 1813)

To investigate the second statement Smeaton tested samples from a large number of limestone deposits, and obtained the best results from a bluish limestone from Wales. On the advice of a chemist he dissolved the limestone in concentrated nitric acid. In his book on the Eddystone Lighthouse he relates that

> the solution appeared very dark and muddy . . . The muddy residuum being brought into an argillaceous state, was very tough and tenacious while soft; and when sufficiently hardened, being worked into a little ball and dried, in that state it appeared to be a very fine compact clay, weighing nearly one-eighth of the original mass. One of the balls having been burnt, became a good, compact brick . . .
>
> From the experiments now related, I was convinced that the most pure limestone was not the best for making concrete, especially for building in water . . . When a limestone is intimately mixed with a proportion of Clay . . . it is made to act more strongly as a cement. (Smeaton 1813, p. 108)

Smeaton's investigation, completed in 1756, led to the first commercial production of modern hydraulic cement, now usually just called cement, in 1796 by James Parker, as he stated in his patent, 'from certain stones of clay or concretions of clay containing veins of calcareous matter' found near London. He called it Roman cement, and the term was widely used for natural hydraulic cements in the nineteenth century.

These naturally occurring deposits were necessarily limited, and in 1811 Joseph Aspdin produced the first cement made from separately occurring deposits of limestone and clay. He called it Portland cement, as Portland stone was the most highly regarded natural stone in London at the time.

J. B. White, the works manager of a competing manufacturer, subsequently claimed to have been the real inventor of modern cement in the 1840s, and this claim has been supported by many historians. However, by 1840 Aspdin had also modified his cement, which consequently became stronger. Records survive of tests made by purchasers of both cements, notably the (London) Metropolitan Board of Works. Skempton (1962–3, pp. 117–52) has examined tests carried out between 1843 and 1871, and came to the conclusion that the two companies produced cements of approximately the same strength. At any rate, the most common type of 'artificial' cement is still called Portland cement in every language.

## Early use of cement and concrete in Australia

Thus the rediscovery of concrete in England, like the production of iron, also occurred in the years just preceding and following the first settlement of Australia.

Natural cement was first exported to Australia in the 1830s or 40s. It was specified for the use in Sydney's first public sewer built from 1855 to 1857 (see p. 158). It was probably used for some of the external stucco work in Melbourne (see p. 57) in the 1860s. In 1854 a Victorian patent was taken out for the manufacture of hollow concrete blocks, and in 1857 a machine was patented for making concrete blocks (Lewis 1988, p. 4).

Natural cement was manufactured near Melbourne in 1862, but the plant closed in 1865. Another factory near Adelaide also did not survive.

Until the 1880s the greater part of the cement used in Australia, and indeed in most other countries, was natural cement, but imports of Portland cement increased, and in 1889 it was produced at Cullen Bullen (later renamed Portland) near Lithgow (NSW). The Cullen Bullen Cement Company was renamed several times, but it continued production.

Also in 1889 the Australian Portland Cement Company was formed in Victoria, and in 1890 David Mitchell (the father of Nellie Melba) established another factory near Melbourne. Mitchell provided some of the finance for John Monash's construction company (see p. 92), and his Emu brand cement was therefore used in most of Monash's buildings.

In 1891 another Portland cement plant was built in New South Wales, and manufacture in South Australia started in 1892. By the turn of the century Australia had five plants, but the bulk of the cement used was still imported.

The cement was first produced in a brick kiln with a chimney on top, somewhat resembling a giant beer bottle, and the clinker was ground and sieved manually. Pictures of early cement manufacture and kilns have been published by Lewis (1988, pp. 4–9) and in the *Constructional Review* (1977, p. 11). Rotary kilns and tube-and-ball mills, which had been developed in the late 1880s, were introduced in 1901 (Lewis 1988, p. 8).

## 'Concrete' houses

A number of houses built in Australia between 1850 and 1880 have been described as concrete houses, However most, as far as can be ascertained, used lime instead of cement.

In southern France, *pisé de terre* was still a common form of construction in the eighteenth century. It is made by ramming loam, a mixture of clay and sand, into wooden formwork. In the early nineteenth century this was improved by an admixture of lime, and this method of construction was publicised in the 1850s by François Coignet (see also p. 87) under the names *béton pisé* and *béton economique*.

The idea was taken up by Herschel Babbage, the son of Charles Babbage, the pioneer of mechanised computing, who came to South Australia in 1851; he became the Chief Engineer of the Adelaide City to Port Railway in 1853, and also one of the founders of the Philosophical Society. In a lecture to the Society on 'Economic Building', reported in the *South Australian Register* on 30 March 1863, Babbage described

> a cheap concrete, as used by Mons. Coignet, a French engineer. He commenced operations with the following compositions for the foundations and the arched roof:- One part of North Adelaide lime and seven parts of gravel and sand taken from the bed of a creek; and for the walls he used seven parts of sand and gravel, three parts of common earth, with one part of unslaked lime. This was the Coignet concrete of which he had read in the *Register*. Some mishaps accompanied the work, but they were traced in each case to avoidable errors, and the experiment established to Mr. Babbage's satisfaction the value of concrete as an economical material.

In a letter to the London journal *The Builder*, published on 6 October 1860, Babbage wrote:

> Settlers at a distance from the town, who find it difficult to procure skilled workmen, I strongly recommend this economic building; and even for stations in the Bush, when, as is often the case, limestone can be found. Concrete buildings would be nearly as cheap as log huts, whilst being infinitely more comfortable, they would be fireproof, set at defiance all attempts by the blacks to burn them, and thus, in case of an attack, enable the inmates to hold out until help arrives.

Babbage presented several more lectures on his 'concrete' houses, the last a year before his death in 1878.

Houses of more conventional type with timber roofs, but also using lime concrete for the walls and foundations, were built by a number of people in South Australia and Victoria between 1860 and 1880. After 1870 some houses were built with site-cast natural cement concrete, but the main use of concrete for housing, except for foundations and floors, has been in the form of concrete blocks and, after World War II, prefabricated panels (see p. 101.).

## Plain concrete

The spectacular use that the Romans had made of unreinforced concrete in arches, vaults and domes was not revived, although there were a few notable exceptions, such as a plain concrete bridge built in Germany in 1904 with a span of 65 metres. The reason was the rapid development of structural iron in the early nineteenth century. In Europe and America long-spanning bridges and the great vaults built for the new railway stations were now constructed with iron. In Australia at that time wide rivers, which might have been bridged in Europe, were crossed by punt or ferry.

Plain concrete was thus not a conspicuous material. Cement was used in foundations and floors, and extensively for the sewers built for the new sanitation systems of the nineteenth century (see p. 158). However, the wider use of concrete in the superstructure did not occur until reinforced concrete was developed.

The age of plain concrete lasted longest in Britain, which had invented modern cement, but was slow to adopt reinforced concrete. In Australia it was much shorter, because it was a relatively late user of cement, but an early user of concrete reinforcement.

In Australia cement was increasingly substituted for lime in the foundations, and it was used for floors and sewers, as it had been in England. Its first use in the superstructure of a building was in the Department of Lands in Sydney in 1876 (Figure 32). The building was designed by the Colonial Architect, James Barnet, but the use of concrete was suggested by the building contractor, John Young (see also pp. 56 and 119).

H. C. Kent, who worked for Young, gave his recollections half a century later:

> I suppose the present day [1924] may be considered the beginning of the Concrete Age as far as building is concerned, for in the 'seventies concrete for constructional purposes was very little used . . .

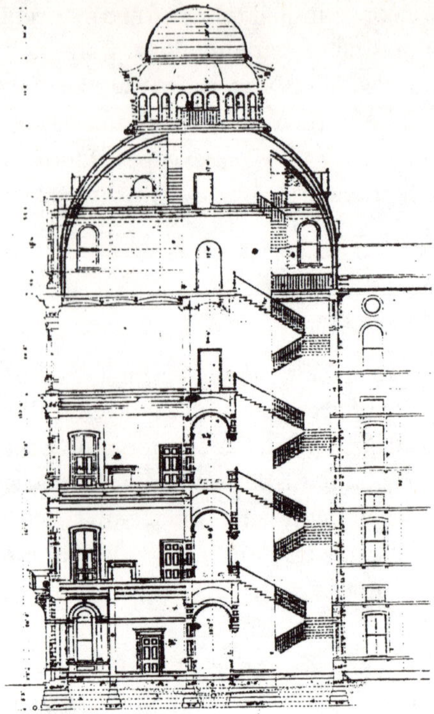

**Figure 32** Section through part of the Lands Department Building, including the Observatory Dome and its Drum, The *Illustrated Sydney News* of 19 August 1876 described it as

> a copper dome 55 feet [17 m] square at the base—changing to an octagonal lantern with a revolving copper dome roof, fitted for an observatory, with a base for the instrument . . . The staircases, of which there will be four, will be of iron, and in each staircase well there will be a patent lift, to be worked by water power. In the internal fittings of the building all the best modern arrangements for heating, light, ventilation, and for communication by speaking tubes, pneumatic bells, etc., will find a place.

(Reproduced by courtesy of the NSW Department of Public Works from a contemporary drawing)

At the Lands Office we used coke concrete for the ceiling vaults over the arcades, and I well remember having some trouble in setting out the centering for the irregular groined vaults of the North End bays . . .

Mr. Young also induced Mr. Barnet to allow him to construct the drum of the Observatory Dome at the Land Office [Figure 32] of concrete, as owing to the openings and the lack of bond it would have been difficult to make a well-tied job in stone or brickwork.

Apart from the above, however, we did not meet any concrete in construction, excepting in our own design, where Mr. Young used it freely.

> He was sometimes called the Father of Bowls owing to the part he played in establishing that here, but I think he might more appropriately have been called the Father of Concrete in our State, for I think he was the first to use it for constructive purposes, . . . and even reinforcing it without any of the present scientific knowledge of reinforcing.
>
> I remember well in connection with a building in Park Street with concrete foundations in which he decided to have a cellar under one portion that we first dug our trenches some 10 feet [3 m] deep and filled them with concrete, and later when it had set, excavated the internal area which had served as our casing. While the work was in progress, Mr. Mansfield, who happened to be passing, looked in and asked me why we were excavating such deep trenches, and when I explained that we were going to take the earth out later for a cellar, he exclaimed 'You must have great faith in the stuff; I wouldn't like to build on it after the earth is removed,' to which I replied: 'It wouldn't be much good, Sir, if we had to rely on the earth to hold it up.'
>
> I mention this as showing how little our leading Architects then recognized concrete for constructive purposes.
>
> In the same building we decided to have a fireproof concrete floor at the First Floor level, and this was reinforced with old railway Barlow rails, which we would now consider ridiculously strong for the purpose . . .
>
> I sometimes smile as I look back and recall our primitive reinforcing of concrete in those days as compared with the scientific reinforcing of today. (Kent 1924)

In the 1880s plain concrete supported by arched sheets of corrugated iron (see Figure 28) was used for fire-resistant floor structures, and in 1895 the first reinforced concrete structure was built.

## The development of the use of reinforcement

There is a surprisingly long interval of more than half a century between the rediscovery of concrete and the invention of reinforced concrete; surprising, because the concept of reinforcement to resist tension in structures of natural stone and brick was well established.

The structures built during the era of prosperity of the Roman Empire were so massive that the horizontal thrust created by the vaults, arches and domes was generally absorbed within the masonry. In Byzantine architecture (which is the late Roman architecture in the Eastern

Empire), the thrust was often absorbed by external wooden or iron ties. This concept was taken over and refined by Islamic architects, and a number of Italian Romanesque builders copied from them the use of external ties to absorb the thrust.

The first great structure of the Italian Renaissance, Brunelleschi's Dome in Florence, had both timber and iron ties embedded in it, a concept that Brunelleschi explained in a memorandum to the Works Office of the Woollen Guild (Prager & Scaglia 1970, pp. 32–42). Iron chains were used to absorb the thrust of the roof structures of Michelangelo's St Peter's Basilica in Rome and of Christopher Wren's St Paul's Cathedral in London (Cowan 1977, pp. 178–202).

In the church of St Geneviève in Paris, later renamed the Panthéon by the Revolution, the architect Jacques-Germain Soufflot used reinforcement in slots in the masonry of the roof that looks not unlike the reinforcement in a modern concrete structure (Cowan 1981, p. 2). The building was started in 1755, the year that Smeaton commenced his investigation on hydraulic cement, and finished in 1781.

It is therefore surprising that it did not occur immediately to somebody that it was so much easier to put reinforcement into fluid concrete than to wrap it around a brick dome, or to chisel slots into the stones of a shallow masonry dome.

## The invention of reinforced concrete

Few people dispute that Smeaton discovered hydraulic cement in 1756; but there is no agreement on the date of the invention of reinforced concrete, or the name of the inventor.

In 1854 a Newcastle-upon-Tyne building contractor, William B. Wilkinson, registered a patent (Joyce M. Brown 1966–7) that would have made him the discoverer of reinforced concrete if anybody, other than his own firm, had made any use of it (Figure 33). Wilkinson bought old colliery ropes and draped them so that they were near the top over the supports and near the bottom at midspan. In a letter that he published in *The Builder* on 5 January 1884 he described some of the experiments he conducted before registering his patent (Brown 1966–7), from which it is clear that he appreciated that in a continuous concrete structure the tension occurred near the bottom at mid-span, but near the top at the supports. It is surprising that this ingenious and economical system did not establish itself, because Wilkinson lived to be eighty-three, and his successful firm survived him. In 1955 a contractor noticed that the floor structure of a building he was demolishing was unusual, and he called in Professor Fisher Cassie from the University of Newcastle. Several more Wilkinson buildings were then located, and the patent was found.

Curiously Britain, which had pioneered plain concrete and iron construction, made no further contribution to the invention of reinforced concrete. The main developments occurred in France, Germany and the USA.

Each of three mid-century patents has been declared by its protagonists as the invention of reinforced concrete. In 1855 Joseph Lambot patented a boat he had built in 1850 from cement mortar and iron bars. François Coignet (see also p. 82) patented *Beton Agglomeré*, also in 1855. This was a concrete rammed into forms; reinforcing bars were not mentioned. Joseph Monier started experimenting with reinforced cement flowerpots in the late 1850s, and obtained a patent for them in 1865.

None of these patents related to reinforced concrete for building construction. However, two provided the basis for later patents. Coignet introduced reinforcement into his concrete. Monier took out additional

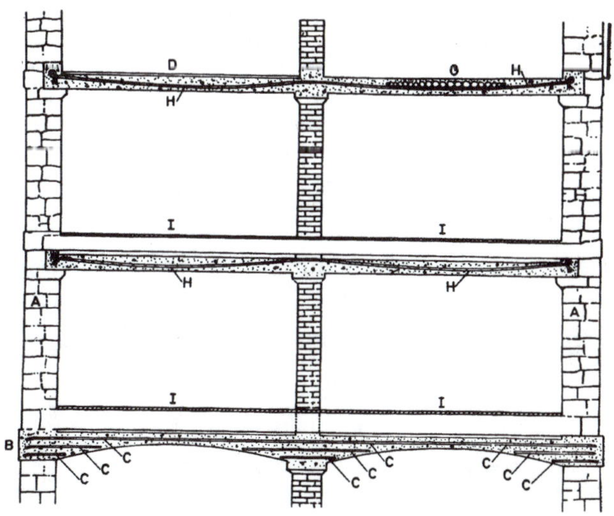


Figure 33 Illustration of W. B. Wilkinson's patent specification, No. 2293 of 1854. The wire ropes (H) are draped so that they provide reinforcement near the top over the supports and near the bottom at mid-span. The drain pipes (G) may be cast in to reduce the weight of the structure. The arched beams at the bottom illustrate an alternative proposal with hoop-iron reinforcement, which Wilkinson does not seem to have used in his buildings.

patents for pipes, beams, vaults, and bridges. He also registered his patents in Germany, Austria–Hungary, Belgium, Spain, the USA, and in Britain and its colonies.

The original Monier patents specified small rods of iron, tied together with annealed iron wire to form a mesh for the flowerpots and their derivatives, the Monier pipes. Mesh reinforcement was also employed for floors. For the larger spans shallow arches were used, which bore a marked resemblance to the plain concrete arches supported on iron sheets that were in common use in England in the mid-century, and were used in Australia by the 1880s (Figure 28); but reinforcing bars were now available to resist any tension that might develop because the arches were too shallow, or because the loading was not uniform (Figure 34). It had been known for more than a century that masonry arches were stable if the resultant thrust lay within the section of the arch; by adding reinforcement, the resultant could now fall a little outside the section of the arch.

The German construction company Freytag & Heidschuch acquired an interest in the Monier patents in 1884. It ceded it to Gustav Adolf Wayss in 1885, and in 1888 the two merged to form Wayss & Freytag.

It is doubtful that Monier ever understood properly what he had patented. He started as a gardener, and there are a number of reports that he was to the end of his life unclear where the reinforcement should be placed in the concrete; to be safe, he preferred it in the middle (Schlüter 1942). Even French contributors agreed at a centenary celebration in 1950 that the proper development of the Monier System was due to Wayss, who was an experienced engineer.

Wayss persuaded two public service engineers, Mathias Koenen and Emil Mörsch, to join the company. In 1886 Koenen published in his book, *Das System Monier*, the first theory of reinforced concrete design, which differed only in details from that subsequently adopted world-

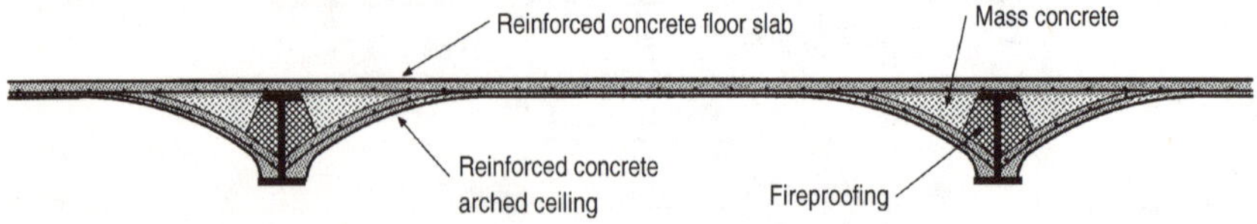


**Figure 34** Arched reinforced concrete floor spanning between rolled steel joists, one of the structures forming part of the original Monier System.

(Redrawn by Peter R. Smith from an illustration in Marsh [1904, p. 86])

wide and used until the 1950s. Mörsch initiated the first rigorous testing of reinforced concrete.

Wayss & Freytag also provided the initial support for a major research programme under Professor Carl von Bach, and later Professor Otto Graf, which continued for several decades, and established the mechanical behaviour and strength of reinforced concrete (Graf 1951).

The acceptance of reinforced concrete was much slower in Britain than on the Continent and in America. There were construction companies that built reinforced concrete structures, but many leading engineers expressed doubt about their durability. British architects were more sympathetic to the new material, and in 1906 the Royal Institute of British Architects set up a Joint Reinforced Concrete Committee, which published a report with recommendations for construction (RIBA 1911). The Institution of Civil Engineers declined the invitation to join the Committee, and set up its own committee, which issued two reports running to 450 pages (ICE 1913). The Conclusion of the Final Report implied that, while reinforced concrete might be acceptable on the Continent, British engineers would be well advised to use more durable materials, and it made *no* recommendations for its construction.

In the USA the development of reinforced concrete was also slower because of its distance from Europe. The first buildings date from the 1870s, and the first patent to an American, Thaddeus Hyatt, was granted in 1878. Thereafter progress was rapid. The experimental work by Professor A. N. Talbot at the University of Illinois soon rivalled that at the University of Stuttgart, and the world's first tall reinforced concrete building was the sixteen-storey Ingalls Building in Cincinatti, designed and built by Ernest Ransome in 1902, a functional building in the tradition of the Chicago School.

The concept of reinforced concrete as such was not patented; but at the turn of the century there were numerous patented systems of reinforced concrete construction. The first English-language textbook (Marsh 1904) describes 42 patent systems, of which 14 originated in France, 9 in Germany, 8 in the USA (most recently registered), 3 in Austria–Hungary, 3 in Switzerland, 2 in Great Britain, 2 in Italy, and 1 in the Netherlands. Some of these were so complicated that they were hardly practical. The Matrai system, devised by a professor at the Budapest Technical University, provided for the reinforcement to follow the lines of principal stress to minimise the amount of iron required, but the work involved could not possibly have compensated for the saving in material, even with nineteenth-century wages. Others bore a close resemblance to modern practice, such as the Monier System as modified by Wayss, Koenen, and Mörsch (Figure 35).

Nor were the systems of equal significance. In the late nineteenth century the Monier System was widely used because of the technical back-up service provided by Wayss & Freytag. For the same reason the Hennebique System, a latecomer registered in 1892, was favoured in New Zealand and in Britain.

## Reinforced concrete comes to Australia

The first reinforced concrete structure in Australia, which received a Plaque from the Engineering Heritage Committee of the Institution of Engineers, Australia, in 1993, was an aqueduct in Annandale (an inner Sydney suburb), spanning 75 feet (23 m), built in 1895, and still in use (illustrated in Cowan 1994, p. 71). It was one of the longest-spanning reinforced concrete structures when it was built, and this was one year before the construction of the first British listed reinforced concrete structure, a flour mill in Swansea (Wales), which was demolished in 1982 after a public enquiry removed the listing.

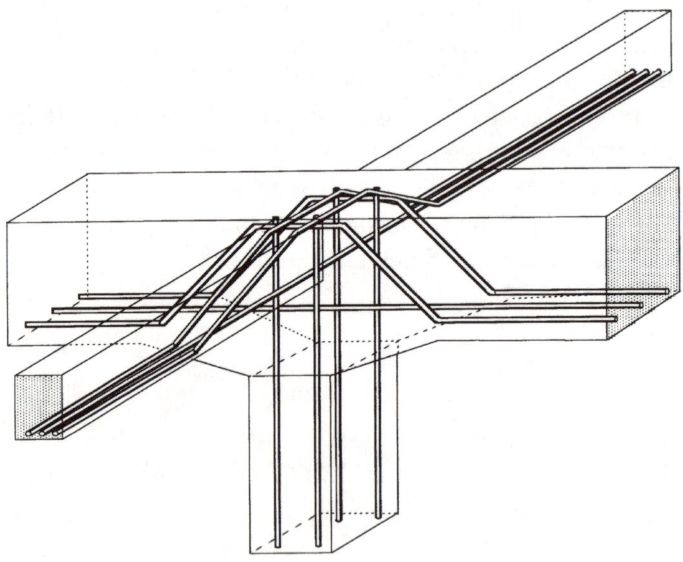

**Figure 35** Wayss's modification of the Monier System, which is already similar in conception to modern reinforced concrete.

(Redrawn by Peter R. Smith from an illustration in Marsh [1904, p. 86])

The aqueduct had originally been designed in brick and plain concrete. The initiative to use reinforced concrete came from W. J. Baltzer, a German-trained engineering draughtsman with the NSW Public Works Department, who in 1895 had produced a typewritten book on reinforced concrete, claimed to be the first in English; the Archives of NSW State Rail have a copy. He persuaded the Engineer-in-Chief for Sewerage (Mitchell 1922) to allow Carter Gummow & Co., who held the local licence for the Monier System, to submit an alternative tender, subject to an unusually long maintenance period of three years. As their tender to construct the aqueduct in reinforced concrete was the lowest, they were awarded the contract. There were subsequently allegations of financial irregularities, and in 1897 the Legislative Assembly set up a Royal Commission, whose report runs to more than 1000 pages (NSW Legislative Assembly 1898). It includes details of the design and construction, and biographical details of the people involved.

Since the principal load on an aqueduct is necessarily uniform, the reinforcement in the arches is probably entirely in compression. The design was in accordance with the original concept of the Monier patent, and with the spirit of the time, which still considered the arch as the appropriate form for a masonry span (see Figure 34). This changed rapidly as the nature of reinforced concrete came to better understood (see Figure 35).

In 1896 Baltzer brought to Australia the first patent machine for making Monier pipes; these were widely used for many years. Gerald W. Mitchell, at that time the NSW Director-General of Public Works, stated that

> the design of the pipes was . . . after numerous tests made the object of fresh patents. These were the basis of the Monier pipes of today, and it may be mentioned the scientific design patented in this country was embodied in patents some two or three years later in the United States and in Germany, two countries in which a patent had value; a fact that showed that in this section of the use of reinforced concrete, scientifically Australia was ahead of other countries. (Mitchell 1922)

F. M. Gummow, an engineering graduate of the University of Melbourne, was fluent in German, and thus had access to the latest information available from the German patentees of the Monier system. In 1897 he submitted, in association with the Melbourne engineers John Monash and Noble Anderson, an alternative tender for the Anderson

Street Bridge over the Yarra River in Melbourne, which had originally been designed as a three-span steel girder bridge, with a maximum span of 100 feet (30 m). His price being the lowest, he was awarded the contract.

In 1897 John Monash, the son of German–Jewish migrants and fluent in German, obtained the Victorian and South Australian Monier licence. In 1905 he founded the Reinforced Concrete & Monier Pipe Construction Company, in partnership with David Mitchell, the manufacturer of Emu Portland Cement (see p. 81), and in 1906 he established the South Australian Reinforced Concrete Company.

In the following years Monash built a substantial number of bridges and city buildings. Although several other patent systems were introduced into Australia (Lewis 1985, pp. 85–90), the Monier companies obtained the majority of the contracts.

George Taylor, the editor of the influential *Building* magazine, started a campaign against the dominance of Monash. The issue of May 1908 carried an editorial surmounted by three headlines: 'A Grave Danger', 'The Growth of a Great Building Combine', 'Its Uninterrupted March across the States'. Taylor then explained that Melbourne had become 'the seat of a combine that used "reinforced concrete" as a lever to undermine Australian building interests. It has at its head a smart business man, who is a capable engineer and a gentleman of high social standing (Major Monash, MCE LLB)'.

Monash 'unostentatiously advised architects to innocently prime cost any reinforced concrete work they had in view, so that the builders . . . had to pay Monash's Company to do it for them at the Company's price'.

Another method used by him was to persuade the client to require 'the tenderer to submit his own design, giving stress diagrams, and describing generally the materials he intends to use . . . As only the Reinforced Concrete Company could tender under those conditions, they practically got the contract without competition'. This was an admission that Monash was the only contractor with a proper knowledge of reinforced concrete.

The dispute came to a head over the decision in 1906 of the Trustees of the Melbourne Public Library to erect a new building (Saunders 1959). The Librarian, Edmund Armstrong, suggested an iron dome, a slightly smaller version of Sydney Smirke's dome over the British Museum Reading Room. The architect, N. G. Peebles, of Bates, Peebles & Smart, consulted Monash, whom he considered the best structural engineer of his day. Monash said he could build it more economically in reinforced concrete (Figure 36), and Peebles made the dome a prime-cost item, which meant that the contract would automatically go to him. Monash's preliminary drawings, dated May 1908, still exist (Saunders 1959, p. 41, and Lewis 1985, pp. 85–90).

A number of contractors, strongly supported by George Taylor, decided to challenge the use of a prime-cost item for this prestigious contract. The Trustees of the Public Library were subjected to political pressure, and when a delegation from the Master Builders' Association formally called on them in 1909, and asked that the specification be rewritten to provide for open tendering, they gave in without giving Monash an opportunity to put his case. Monash submitted a tender, but

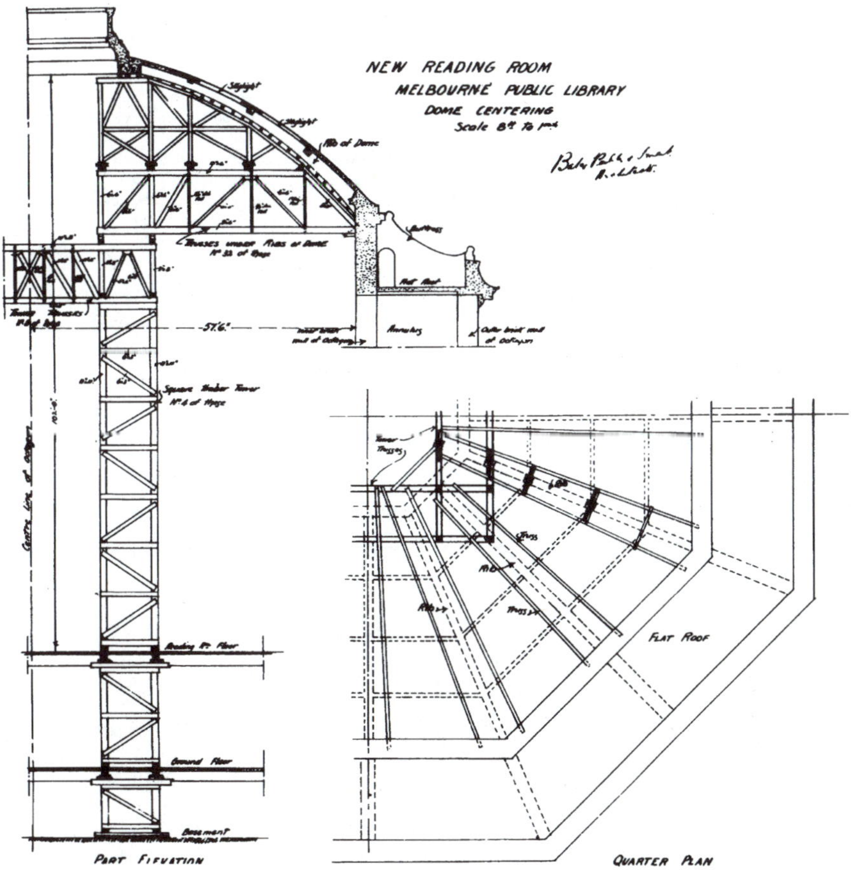


**Figure 36** Formwork for the 115 foot (35 m) diameter dome of the Melbourne Public Library, 1911.
(Reproduced from Saunders [1959, p. 44])

it was not the lowest. The contract went to the Swanson Brothers, who brought in the Trussed Concrete Steel Company of London, who had just built the domed Central Hall in Westminster, as a subcontractor for the dome. However, the design is very similar to that shown in Monash's drawings, except that the reinforcement was changed to the American Truscon Company's patented Kahn bars (Figure 37). The design should still be credited to Monash.

The dome of the Melbourne Public Library created a new world record for a reinforced concrete dome by one foot, when it was completed in 1911, and it held it for a year. Perhaps more significant is that it is still the longest-spanning reinforced concrete dome in Australia.

In 1909 Monash claimed that certain of his patents had been infringed. In an editorial in the May 1909 issue of *Building* Taylor wrote that he was bluffing, and challenged him to go to law. Monash, who had a degree in law, must have judged that Australian courts could not be relied on to protect foreign patents; as mentioned previously in this section, Gerald Mitchell held the same view a few years later. When Monash failed to act, Taylor announced in 1910 that reinforced concrete construction was now open to all (Serle 1982, p. 186). In 1919 the German and Austrian patents were declared void by the peace treaties, and this spelt the end of patent systems worldwide; but in Australia they had effectively come to an end in 1910.

Unlike the previous innovations in Australian construction, reinforced concrete was not introduced from Britain, but from the European Continent and from America. Monash was conscious of the problems this raised, and he asked his brother-in-law, Walter Rosenhain, who had become Superintendent of Metallurgy at the (British) National

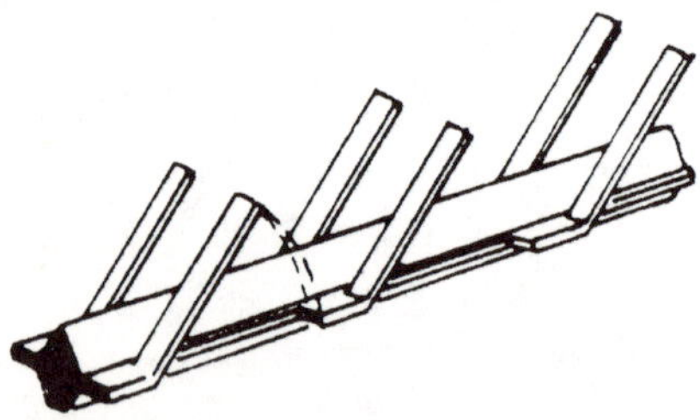

**Figure 37** The patented Kahn bar, a rolled section with flanges on both sides, which were split and bent up as shear reinforcement. The bar was invented by Albert Kahn, who is best known as the designer of a number of industrial buildings, mainly in steel, such as the Chrysler Assembly Plant near Detroit.

(Reproduced from Jones [1913, p. 223]).

Physical Laboratory, to keep him supplied with details of any new reinforced-concrete work in England, because 'nothing talks so strongly with the profession and with State Departments in this country than to be able to quote English practice. America is regarded as unreliable, while anything savouring of being "Made in Germany" is unwelcome as a precedent' (Serle 1982, p. 165). However, most of the technical information used during the introduction of reinforced concrete into Australia came from Germany, and the first Australian Concrete Code was based on the American code.

Monash's success in the years preceding World War I was not primarily due to the fact that he held the Monier patents, but to his clear conception of the potential of reinforced concrete as a new structural material. Monash explained how 'the compressive strains were taken by the concrete and the tensile strains by the iron'. He claimed that in Australia it was a much cheaper construction than iron and steel, it cost about the same as timber, but it was fireproof, and concrete was 'a fine plastic substance which could be moulded to any form of beauty'.

Monash had been a part-time soldier since his days as a university student, and he had reached the rank of Colonel in the militia when the war broke out. Towards its end he became the Commander-in-Chief of the Australian Army. He did not return to structural engineering after the war, but as Chairman of the Victorian State Electricity Commission he developed the production of electricity from brown coal at Yallourn (see p. 122).

Reinforced concrete structures required a more scientific approach than those using traditional materials or iron, where generous sizing normally ensured structural safety; it was also necessary to place the reinforcement where it was needed. A common cause of structural damage in the early days was failure to provide reinforcement to resist negative bending moments or shear forces.

The newly established engineering school in the University of Sydney took a keen interest in the new materials. Professor W. H. Warren had given evidence in late 1987 to the Royal Commission on the Annandale Aqueduct referred to above (Fraser 1985, p. 85). It seems that the following year he lectured on reinforced concrete, because an optional question was included in the Materials and Structures III examination paper for 1898, and further questions appeared in the examinations in each subsequent year.

Warren conducted tests, some supported by Gummow Forrest & Co., to supplement those by Bach and Graf at the University of Stuttgart

(Warren 1902, 1904 and 1905), which included strain measurements with a Martens mirror extensometer, and he included 126 pages on concrete and reinforced concrete theory in the second edition of his textbook (Warren 1921).

## Some other early reinforced concrete structures

Concrete floors supported on arched corrugated sheets (Figure 28) were probably first used in Sydney's Corn Exchange in Sussex Street in 1887. Concrete floors with expanded metal were first used in Challis House in Martin Place in 1890 (most of this building, including the original floors, has since been replaced by a new building of the same name). Both these methods of floor construction had become common by the turn of the century, replacing timber floor construction, until they were in turn replaced by reinforced concrete floors of the type still used today (see p. 100).

The 'flat slab' floor was invented independently by Robert Maillart in Switzerland, whose work had no influence on Australia, and by C. A. P. Turner in America. The floor slab is supported directly on the columns, through enlarged column heads, thus doing away with the beams normally used to support a floor. Turner registered his patent in Australia in 1906, but he did not erect his first building in St Paul (Minnesota) until 1907. The first flat-slab building in Australia, and the first outside America, was a cigarette factory built in 1908 in Drewery Lane, just across the road from the Melbourne Public Library, designed by the engineer H. R. Crawford. It had octagonal columns and used the then popular but now obsolete 'four-way' system (illustrated in Lewis 1988, p. 17).

An even more remarkable structure was the Dennys Lascelles Woolstore in Geelong (Victoria) (illustrated in Lewis 1988, p. 19), built in 1910 by the engineer E. G. Stone, and demolished in the 1980s. Its roof had trusses with arched top chords with a span of 175 feet (53 m); it remained the longest reinforced concrete span in an Australian building for many years. It made use of the French Considère System, in which the compression members had spiral reinforcement that restrained the lateral expansion of the concrete, and thus increased its load-bearing capacity (see Figure 39).

The Rotunda of Newman College at the University of Melbourne has one of the few doubly curved concrete structures of that era (Figure 38). It houses the dining room, and the ribbed reinforced concrete vault creates an impressive interior space. Walter Burley Griffin had won the competition for the planning of Canberra in 1912, with the right to private practice. He kept two separate offices, one in Melbourne for Canberra, and another for his private practice in Sydney, which produced the design for Newman College in 1915 in association with Augustus Fritsch.

Lewis (1985, p. 86) considers that the administration building erected by Monash in 1905 for Mitchell's cement company and his own construction company was the first with a reinforced concrete frame (illustrated in Lewis 1988, p. 10). It has only windows and spandrel beams between the 'columns', but these columns were wide and could be considered narrow load-bearing walls. It was one of the few buildings in the first half of the twentieth century with off-the-form concrete on the outside.

The first 'conventional' reinforced concrete frames appear to have been used in two buildings erected in 1918 adjacent to one another in North Terrace, Adelaide, the Liberal Club and Shell House. Both are eight storeys high, and both have masonry facades.

**Figure 38** Cross section through the rotunda of Newman College, University of Melbourne, designed by Walter Burley Griffin in 1915. The surmounting reinforced concrete spires and pinnacles have been removed.

(Reproduced from 'Walter Burley Griffin', *Constructional Review* [1977, p. 28])

## Reinforced concrete after World War I

The era of interesting reinforced concrete buildings in world class came to an end with the 1910s, partly because the most economical uses of the new material had been established by the experimentation of the two previous decades, and partly because with the opening of the Newcastle steel works (see p. 68) steel became readily available at a lower price.

Union House was Sydney's first reinforced-concrete frame building, but it had originally been designed as a thirteen-storey steel frame in 1919. The first floor was built with steel columns, but as there was a shortage of steel in the immediate post-war era, it was redesigned by Charles Reid in reinforced concrete, using Kahn bars. It was demolished in 1977. An editorial in the January 1920 issue of *Building* listed four other reinforced concrete buildings erected in Sydney in 1919, but none were more than a few storeys in height.

In 1921 the thirteen-storey block, Astor Flats, designed by A. S. Macdonald and still standing in Macquarie Street, Sydney, was claimed to be the tallest reinforced concrete frame building in Sydney. Because of the height restriction introduced in 1912 (see p. 76) no significantly taller building was erected subsequently until after 1957 (see p. 188).

In 1925 the fifth floor of the six-storey British Australasian Tobacco Building in Melbourne collapsed during construction, with the loss of four lives. A subsequent investigation established that the concrete mix was very sandy, and that far too much water had been used, so that one beam showed a layer of laitance 4 inches (100 mm) thick. This did considerable damage to the reputation of reinforced concrete, but it also ensured better mix control in subsequent buildings.

One argument against the use of reinforced concrete in multi-storey buildings is the space taken up by the internal columns. In the David Jones Store in Elizabeth Street, Sydney, reinforced concrete was used for the *floors* on all storeys. The *columns* in the four lowest storeys, which carry the heaviest loads, were steel and those in the upper stories were reinforced concrete, using spiral reinforcement (Figure 39). While this is rarely used today, it was popular in the 1920s because the restraining action reduced the size of the columns, at a time when the limiting stresses in steel and in concrete were much lower than they are now.

During the inter-war years structural steel established itself as the normal material for multi-storey buildings above six storeys, although

Figure 39 Contemporary photograph of the David Jones Store, Elizabeth Street, Sydney, under construction in 1926. The reinforcement of the beams and slabs is already similar to that used today, except that the bars are not deformed. The columns on the lower floors are steel, and the columns on the upper floors have spiral reinforcement, which was the essence of the expired Considère patent. The concrete was machine-mixed on the site, and transported in wheelbarrows.

(Reproduced by courtesy of David Jones Ltd)

reinforced concrete buildings were also built, sometimes up to the full height permitted by the building code (150 feet in Sydney).

However, during the 1920s reinforced concrete became the normal material for multi-storey floor construction, partly because reinforced concrete had become cheaper, and partly because large timber sections became more expensive as the virgin forests were cut down faster than the natural replacement rate of the trees. There was just one solitary example of the use of hardwood floors in a woolstore in the 1930s (see p. 36).

Miles J. Dunphy, who became Head Teacher of the Department of Architecture of the Sydney Technical College (the first institution to provide a professional architecture course), gave his reminiscences on 19 May 1977 in a letter to the Dean of Architecture of the University of New South Wales (the letter has been deposited in the library of the

Cement and Concrete Association of Australia). During the 1920s he had worked as chief assistant to A. S. Macdonald, who was one of the first consulting engineers, and the structural designer of the David Jones Store (Figure 39); he later founded the firm of Macdonald Wagner & Priddle, now absorbed in the Connel Wagner Group. Dunphy recalled that these years

> coincided with the great change in building design from the brick shell with hardwood storey posts and beams as internal structure, to a structural steel frame encased in concrete, with combination outer walls of brick with stone or architectural terra cotta facing.

This was made possible by the use of reinforced concrete for the floors.

Welded mesh, which had first been imported from England for use in Melbourne in 1913, was soon locally manufactured, and became the favoured reinforcement for concrete floor slabs.

## Sitecast and precast concrete

It had been known from experience since the early days of concrete that the driest mix made the strongest concrete. Both François Coignet for his *Béton Aggloméré* and Joseph Monier for his flowerpots (see p. 87) specified a mix so dry that it had to be rammed into the formwork. As reinforcement became more complex (see Figures 34 and 35), however, a more fluid mix was needed. The more water was added, the easier it became to place the concrete, and the lower the strength of the concrete. Workmen were naturally tempted to do that; it was the main cause of the collapse of the B.A. Tobacco Building in 1925 (see p. 98).

In 1919 Duff Abrams established from experiments in America that the strength of concrete was inversely proportional to the ratio of water to cement (Abrams 1919), and it then became possible to determine scientifically how much water should be used in the concrete mix to achieve a specified strength.

Concrete was at first mixed manually with shovels. Concrete mixers were imported into Australia in the 1870s, but they became common only after the turn of the century. Manual mixing was still used as late as 1911 for the dome of the Melbourne Public Library (see p. 92).

Ready-mixed concrete, which removed the concrete mixer from the building site to a central mixing plant, was invented in the USA in 1926. The concrete was initially transported in a special horizontal agitator, which could be tipped, known as a 'tumble bug'. The first Australian ready-mix plant was built on Glebe Island in Sydney in 1939 (*Constructional Review* 1977, p. 53). By the mid-1950s ready-mix concrete had displaced concrete mixers from most metropolitan building sites in Australia.

The ancient Romans always used permanent formwork of brick or stone for their concrete structures, except for the underside of vaults and domes, where permanent formwork of brick or stone would have been impracticable, so that the concrete was left off the form.

This sensible rule was revived when concrete was reintroduced in the nineteenth century, and off-the-form concrete was only rarely used on the exterior surfaces of buildings in Australia until the 1970s. One of the few examples of its use on building exteriors before World War II was Winthrop Hall at the University of Western Australia, designed by the Melbourne architect Rodney Alsop in 1927; the formwork was carefully detailed and the surface weathered well.

In the early twentieth century precast concrete was used far less than site-cast concrete. An early example, and a well-known landmark, is the lighthouse at Bradley's Head in Sydney Harbour, built by Carter Gummow & Co. in 1904 (Max Dupain's photograph of the lighthouse is reproduced by Lewis 1988, p. 76).

After World War II there was a serious shortage of housing, and precast concrete that could be mass produced became more common. A number of precasting works were built in various parts of Australia, the biggest by the Victorian Housing Commission, which acquired for £1 a redundant car assembly plant that had been used for defence equipment, to make precast concrete wall and floor panels. The concreting and handling operations and the transport were highly mechanised, and the panels were assembled by bolting. The single-storey houses, however, were no cheaper than conventional brick-veneer houses.

The Housing Commission then built three-storey walk-up apartment buildings, and eventually high-rise apartments with lifts. A thirty-storey block, assembled from precast load-bearing concrete panels by bolting and prestressing without a frame, is the tallest ever built that way (illustrated by Lewis 1988, p. 106).

The factory was closed after about fifteen years of operation because the buildings, although comparable to the best overseas, were neither cheaper than conventional construction, nor were they popular (see p. 194).

However, precast concrete panels were widely used from the 1950s to the 1970s for the curtain walls of commercial multi-storey buildings, supported by a skeleton frame (see p. 200).

## Prestressing and flat-plate construction

The concept of prestressed concrete is almost as old as that of reinforced concrete (Figure 40). In reinforced concrete the fine tension cracks are bridged by reinforcement that resists the tension; prestressing induces compression in the concrete before the load is applied, so that the tension does not develop. In 1886 C. W. Doehring took out a German patent, and in 1888 P. H. Jackson patented a method in America. Their test beams carried a satisfactory load immediately after the prestress had been applied, but in both systems the prestress vanished after a few days.

This was the age of Puccini's *La Bohème* and of novels in which romantic heroines died of consumption, and the phenomenon was called concrete tuberculosis, for which there seemed to be no cure. It was not until the shrinkage and creep of concrete were understood, that it could be explained.

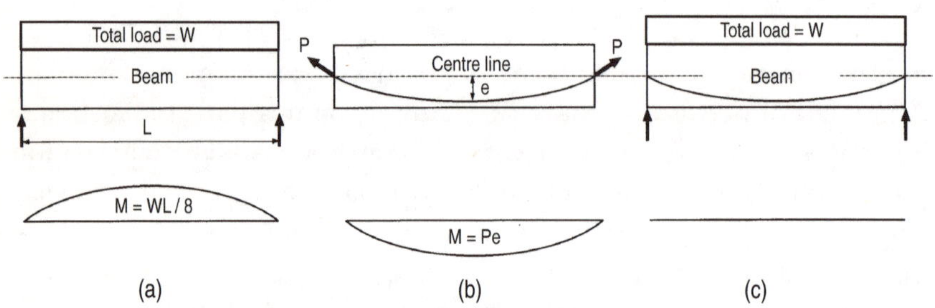


**Figure 40** The principle of prestressed concrete:
(a) A loaded beam is subject to a positive bending moment that varies parabolically, as explained in any book on structures (e.g. Cowan & Wilson 1981, p. 59).
(b) An equal and opposite bending moment can be induced by a prestressed cable whose eccentricity varies parabolically.
(c) The net result is a beam that is entirely in compression. As the bending is eliminated, the tension in the concrete is also eliminated.

In practice allowance must be made for the variation in the load. A beam might be designed for the full weight of the building, which is the greater part of the total load, plus half the contents of the building. The beam then has a small negative bending moment when the building is empty, and a small positive moment when it is fully loaded.

(Drawing by Peter R. Smith)

The prestressing steel extends elastically as it is prestressed. Shrinkage is caused by the chemical reaction between cement and water, and it results in a contraction of the concrete. Creep under the action of the prestressing force causes a further contraction. These produce an equal elastic contraction of the tensioned steel, and with the normal reinforcing steel available in the 1880s the entire prestress was lost by this contraction. In the 1920s Karl Wettstein in Germany and Eugène Freyssinet in France used high-tensile wire with a strength about ten times as high as the steel bars in the 1880s. It extended about ten times as much when it was prestressed, so that the loss of prestress became acceptable.

By 1939 prestressed concrete had not achieved much acceptance outside France and Germany. In 1940, after the fall of France, the Germans used it for extensive defence structures along the Atlantic coast. Steel was in short supply: reinforced concrete needed less steel than structural steel, but prestressed concrete needed very much less. After the defeat of Germany Allied investigating teams reported on these prestressed concrete structures, and this created an immediate world-wide interest in prestressed concrete.

In Australia some prestressed concrete structures, notably the six-storey ice-tower for the construction of the Warragamba Dam (NSW), were built in the early 1950s. There was then a widespread impression that prestressed concrete, in which the concrete was free from cracks, in due course would replace reinforced concrete, which inevitably had fine cracks under normal working loads.

In the immediate post-war period, when there was a worldwide steel shortage, prestressed concrete again had the advantage of requiring far less steel. When the shortage eased, it became clear that its higher cost would limit its use to complex structures, such as the Sydney Opera House.

Its advantages for use in ordinary buildings were not fully appreciated until the late 1960s when building heights in Sydney started to exceed thirty storeys. Internal columns would have occupied much valuable space, particularly in the lower storeys, and they would have prevented a flexible layout. The floor structure was therefore designed to span from the lift core to the columns on the facade. For these long spans prestressed concrete was much lighter and more economical (see p. 193).

A flat *plate* is another post World War II innovation. The first Australian flat *slab* building was erected (see p. 103) in Melbourne in 1908.

A flat slab is a beamless floor, but with enlarged column heads to facilitate transfer of the load from the slab to the column. In 1940 a New York consulting engineer, Joseph Di Stasio, proposed that the external concrete column head be replaced by an 'internal' one consisting of additional reinforcement. It was then found that this reinforcement could also be omitted if the slab was made a little thicker, which presented no problems in residential and other buildings with relatively small spans.

The flat plate was immediately adopted in Australia, where few multi-storey residential buildings at that time had either built-in heating or air conditioning, so that there was no need for a false ceiling; electric cables could be cast into a thick slab. Lex Blakey (1965), later Chief of the CSIRO Division of Building Research, noted that it had become 'an Australian structural form'. The concrete slab was generally given a sprayed ceiling finish, and the floor was covered with a carpet. It was an economical form of construction, because of the simplicity of the formwork and the relatively small cost of a little more concrete. The extra thickness also improved the sound insulation.

# 5

# THE LUMINOUS ENVIRONMENT

## Daylight

Artificial light was developed very early. Candles or oil lamps have been known for more than three thousand years, but an individual candle or lamp did not provide a great deal of light, and adequate illumination after daylight faded was the privilege of the wealthy. This was not greatly altered when gaslight was introduced, until the Welsbach incandescent mantle came into use in the late 1860s (see p. 115). Before the twentieth century daylight was the most important source of illumination.

Canberra is roughly at the same latitude as Gibraltar, so that Australia is fortunate in having a much smaller variation in the length of daylight between summer and winter than England. Daylight therefore sufficed for many activities in the nineteenth century.

Additional daylight was provided by using large areas of glass; the prototype for this form of construction was the Crystal Palace built in London in 1851. This, in turn, was derived from the French orangeries of the seventeenth century, made possible by the development in France of new processes for making larger sheets of glass.

Orange trees were introduced into Europe in the late Middle Ages. The high esteem in which they were held was due to the scent of their blossoms, rather than their fruit; but in Europe they cannot be grown outdoors, except in the south. In France the trees had to be kept inside an orangery or greenhouse, initially a building with large glass walls. During the eighteenth century it was often built entirely of glass and iron.

Joseph Paxton, the head gardener of the Duke of Devonshire, for whom he had built several large greenhouses, became the designer of the building for the First International Exhibition in London in 1851 largely by accident, after many weeks of inconclusive public debate (Cowan 1978, p. 304). With only eleven months to go to the opening day of the Exhibition, Paxton submitted through an intermediary a design for a large glass-and-iron building, which he claimed could be built in the time available. The exhibition opened on time, and the 'Crystal Palace' caught the public imagination.

It inspired numerous glass-roofed structures in London, and these were soon copied in Australia. The first cast-iron-framed glass-roofed shopping arcade was Queen's Arcade, built in Melbourne in 1853 between Little Bourke and Lonsdale Streets (Garside & White 1963, p. 80). It survived for about fifteen years. The Royal Arcade between Bourke and Little Collins Streets was built in 1869 with a cast-iron roof; the annex to Elizabeth Street is a later addition in wrought iron. The arcade is still in use, but there has been a good deal of reconstruction.

The Victoria Arcade, built in Melbourne in 1873, is no more, but the Block Arcade, from Elizabeth to Collins Streets, opened in 1894 (Figure 41), is still in use.

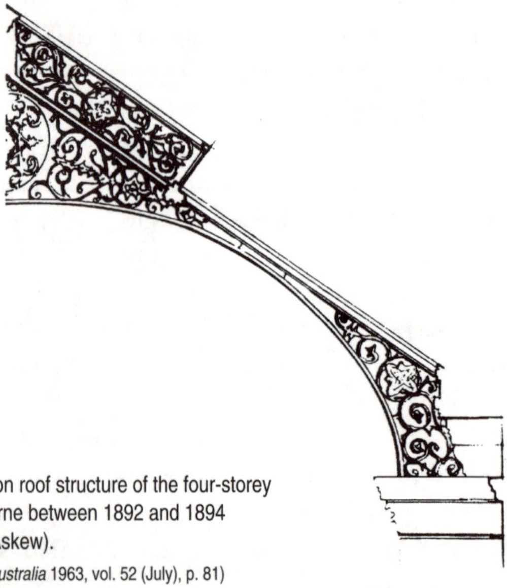

**Figure 41** Part of the cast-iron roof structure of the four-storey Block Arcade, built in Melbourne between 1892 and 1894 (architects: Twentyman and Askew).

(Reproduced from *Architecture in Australia* 1963, vol. 52 (July), p. 81)

Sydney's first arcade was the Sydney Arcade, built in 1881 between King and George Streets (Heath & Moore 1963, p. 85), followed by the Royal Arcade in 1882 between George and Pitt Streets, the Victorian Arcade in 1887 between Castlereagh and Elizabeth Streets, and the Imperial Arcade in 1891 between Pitt and Castlereagh Streets (where there is now an electrically lit arcade). Three of these arcades survived to the 1960s. Now only the Strand Arcade (Figure 42), opened in 1892, is left, and this has had a major reconstruction after a fire.

All the glass arcades built in Australia in the nineteenth century had provision for artificial lighting, the type depending on the time of their construction.

**Figure 42** Strand Arcade, built in Sydney in 1891, a glass-roofed daylit arcade; in addition, it was fitted with electric light from a private power supply (architect: John B. Spencer). The present electric lights were installed during a reconstruction.
(Photograph by Henry J. Cowan)

More arcades were built in the early twentieth century, with some glass roofing for daylight, but with greater reliance on electric light, and generally steel-framed. Arcades then lost favour, and when arcade building resumed after World War II daylight no longer played a significant part in the design.

The restored Queen Victoria Building in Sydney (see p. 71) is a good example of the important part daylight played in the nineteenth and early twentieth century, even though a large part of the illumination now comes from electric light. It has a barrel-vaulted glass roof, and a glass dome over the crossing. The basement was lit by a large number of prismatic glass pavement lights, which have not been restored. The building thus had good daylight for pedestrian circulation, but some of the interior spaces would have needed artificial light, as would the use of the building after sunset; this was originally provided by gas light.

In the nineteenth and early twentieth century most multi-storey office buildings had light wells (Figure 43) which were just large enough to give some illumination to the space adjacent to the interior windows. The Royal Exchange Building in Bligh Street, Sydney, was probably one of the last to be built that way (Heath 1996). The light well sometimes terminated in a glass roof for the ground floor.

**Figure 43** Light well in a nineteenth-century office building, to provide daylight for the interior of the building.

(Drawing by Peter R. Smith)

Until the 1960s many Australian buildings had little artificial lighting. Some schools had electric light only in the corridors for safety, and in a few rooms used by staff or parents in the evening. Many staff rooms in the University of Sydney had no light fittings installed until the 1950s, because it was assumed that they would rarely be used after it was dark.

The same applied to some factories and commercial buildings. The Commonwealth Department of Labour and National Service had an Industrial Welfare Division during and in the years following World War II, which devoted much effort to the improvement of natural light. The international recommendations for the calculation of daylight (CIE 1970) are largely based on its work, one of the few instances where an international standard has been based on work originating in Australia.

As electric lighting improved, the utilisation of daylight declined, and many offices and shops were lit entirely by artificial light. The energy crisis of 1973 (see p. 123) created an interest in reducing electricity consumption by utilising solar energy. Daylight is a *direct* use of solar energy. It can be utilised even when uniform lighting levels are required if it is combined with electric light, which is varied to allow for the change in daylight with sky conditions. This can be done with sensors and automatic controls, and is probably more economical than most other methods of utilising solar energy (see p. 144), which generally require conversion of the solar energy to electricity.

## Candles and lamp oil

Until gas light was introduced (see p. 112) artificial lighting was mostly by candles or oil lamps.

Some candles were made from beeswax. These smelled pleasantly and burned without sputter, but they were in short supply. In medieval times they were generally reserved for the church and the court; in the early days of Australia only the well-to-do could afford to pay for them.

Until the mid-nineteenth century most candles were made from tallow and other animal fats. Candles were originally made by dipping a fibre wick into a tub of molten wax or tallow, allowing it to set, and repeating the operation until the candle was thick enough.

The poorer people in Australia, who could not afford to pay for proper candles or oil lamps, could produce a home-made 'candle' cheaply by holding a piece of rag in a cup or bottle, and pouring dripping around it.

The light would have been dim and the smell unpleasant, but it was easy to make.

Machines for manufacturing candles were developed in Europe in the early nineteenth century. By the 1850s candles were produced in Australia from boiled-down animal carcasses (Shar Jones 1984, p. 13).

The fact that oil oozing out of the ground in various parts of the world was combustible had been known since antiquity. However, it could not be utilised, except in a few places where it was thin enough, until E.W. Binney, a Manchester chemist, distilled it in 1847 to separate the fraction, which is now known as kerosene, from the heavier fraction now called paraffin (Cowan 1978, p. 255). The immediate interest was in the use of kerosene for lamp oil, but after a few years it was found that the paraffin wax made good and cheap candles, and these largely replaced tallow candles.

In antiquity most oil lamps burned vegetable oils, and this continued to be the principal fuel until whale oil became available for lamps in the eighteenth century.

As the number of whales in the Arctic region declined, the whalers, mostly American, moved to the Southern Ocean. In 1788 the first whaleship penetrated into the South Pacific (Clarke 1965). Captain Watkin Tench (1961, p. 298) noted four whaleships had sailed into Sydney Harbour within six weeks in 1791, when Sydney was still a small settlement. More than a dozen whaling stations were established in New South Wales and Tasmania, and until the 1830s whale and seal products were Australia's main export (Bassett 1993, p. 291). Lamp oil was therefore not among the commodities in short supply in the early nineteenth century. However, when kerosene became available, soon after the Victorian gold rush, it largely replaced whale oil.

Some of the kerosene came from Britain, but most was imported from the USA. Some was also produced locally from oil shale (Birmingham 1979, pp. 119–130).

Oil shale was mined in sixteen different locations, mostly in New South Wales; the first mine opened in 1865. The precipitous railway from Katoomba into the Jamieson Valley, built to bring up the shale from the mine in the valley, but now used by tourists, is one of the few remains of the industry. None of the operations was on a large scale, and in the nineteenth century most of the oil shale was exported. However, some was processed locally, and the distillation products included kerosene.

## Chandeliers and lamps

Although the light of a single candle or oil lamp, even of the best quality, is not very bright, a large assembly can give excellent illumination. Alastair Laing (1982, p. 13) states that a candelabra in St Peter's Basilica in Rome had 1365 lights in medieval times. Chandeliers with a hundred candles were not uncommon in the palaces and the houses of the aristocracy in France in the seventeenth century.

Polished pieces of lead crystal, a material developed in the late seventeenth century, were originally added to chandeliers as reflectors, but they were soon turned into elaborate decorations, and crystal 'chandeliers' are still used decoratively as electric light fittings.

There is not a great deal that can be done to improve the performance of individual candles, although a glass chimney helps. However, during the eighteenth century a large number of innovations in the manufacture of oil lamps made them more efficient. One of great influence on subsequent designs was the patent taken out in 1784 by the Swiss scientist Ami Argand, explained in the caption to Figure 44.

The lamps imported into Australia and manufactured locally are described by Peter Cuffley (1973) with the aid of numerous illustrations. During the nineteenth century there were many more changes to

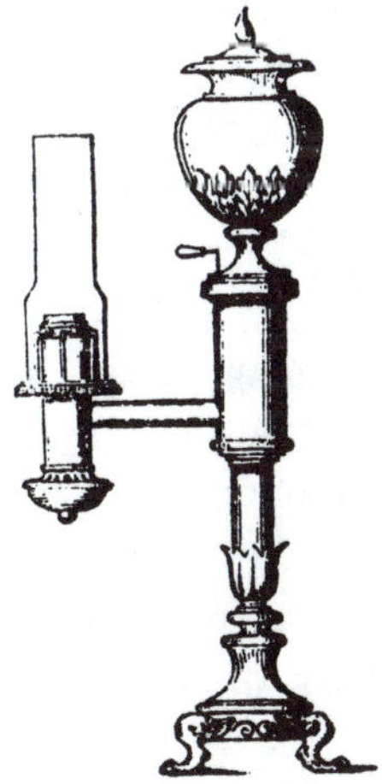

**Figure 44** The Argand oil burner has a circular wick to increase the amount of light. A glass tube protects the flame from currents that might cause it to flicker, and creates an upward draught to ensure a supply of oxygen. For heavy oil the reservoir is located *above* the level of the flame to ensure a constant supply of oil. This gravity feed is not required when the more volatile kerosene is used. (Reproduced from Shar Jones [1984, p. 6])

improve the illumination provided by the lamps, and their safety. This became a bigger problem through the increasing use of kerosene, which burns with a brighter flame than the heavier oils used previously, but it is more volatile, and therefore more liable to explode.

## Gas

It has been known since medieval times that coal in certain circumstances gives off a gas when heated. When coke was made in the early eighteenth century for metallurgical purposes, the gas was at first treated as a waste product, but a few decades later a number of people in various parts of Europe collected some of this coal gas, and utilised it for lighting, heating or cooking. Among them was James Watt (see p. 59) who lit his Birmingham factory with gas. On the occasion of the Peace of Amiens, which ended the first war with Napoleon Bonaparte in 1802, the factory was floodlit with gas. According to a contemporary account (Billington 1982, p. 414) 'all Birmingham poured forth to view the spectacle: and strangers carried to every part of the country an account of what they had seen'.

The idea of using pipes laid under the surface of the street for a public supply was not new. Pipes had been used to supply water in Roman times, and again since the Middle Ages in monasteries and in a few cities (see Figure 65). The proposal for a public supply of *gas* was first raised towards the end of the eighteenth century, and the first gas company commenced service in London in 1814. Its main purpose was to provide street lighting.

Until then the wealthy had their way lit at night by an escort of servants carrying rushes dipped in tallow, which might burn for an hour, while ordinary people tended to arrange their social life so that they could make use of the light of the moon; for example, the Lunar Society, a society of amateur scientists in Birmingham, whose members included Joseph Priestley, James Watt, John Smeaton, Josiah Wedgwood and Erasmus Darwin, took that name because it met on the Monday nearest to the full moon.

In Australia the initiative came from Ralph Mansfield, a Methodist missionary turned businessman, who advocated gaslight 'for the strong protection offered to person and property', 'the pleasure of perambulating the streets thus brilliantly lit', and 'the valuable facility to business, in shops and in offices' (Carroll 1988, p. 43). The Australian Gaslight Company was formed in Sydney in 1836. It hired a British engineer, James Bryan, who arrived in 1839, and the first twenty-two gas lights were lit on Queen Victoria's birthday in 1841. Later in the same year the Royal Victoria Theatre in Pitt Street was lit by gas. The first gasworks were built near Millers Point, at the junction of Kent Street and Gas Lane. By 1846 the pipes extended in the Central Business District from Argyle Street to Campbell Street (Fraser 1989, p. 173).

The gas was produced in horizontal retorts, and bubbled through lime water to remove hydrogen sulphide, which would have given it a smell of rotten eggs. It was then passed through a condenser to liquefy tar and ammonia, which were initially considered useless, and dumped into the harbour. It was stored in a gas holder consisting of a circular tank and a circular bell made of cast-iron plates, which was placed over it with a water seal.

It was assumed at first that if the gasworks were situated at the lowest point, the coal gas, which is lighter than air, would flow through the pipe system of its own accord. However, it was soon found that the storage tank needed to be loaded to apply pressure to the gas.

The initial intention had been to supply only gaslight for use at night; but the retorts needed to operate for twenty-four hours for efficiency, and the daytime production had to be stored. The company's advocacy of the use of gas for cooking and heating was thus initially an attempt to provide a use for the daytime production, and limit the storage capacity required. However, it saved the gas companies from an early demise when electric light became common in the twentieth century.

The earliest gas burners simply had holes in the gas pipe, which produced thin jets of gas. These were replaced by 'fish-tail burners', which remained in use both for street lights and for interior lighting for many years (Figure 45), and by 'Argand' gas burners (Figure 46).

Gas burners had the advantage that there was no wick to be trimmed, and that they could be extinguished easily, but the illumination they provided was not much better than that available from candles and oil lamps, and they also needed a flame to light them. Many people thought they created a worse smell. Lady McLeay (Shar Jones 1984, p. 14) is reported to have preferred the use of chandeliers in Elizabeth Bay House (see p. 45).

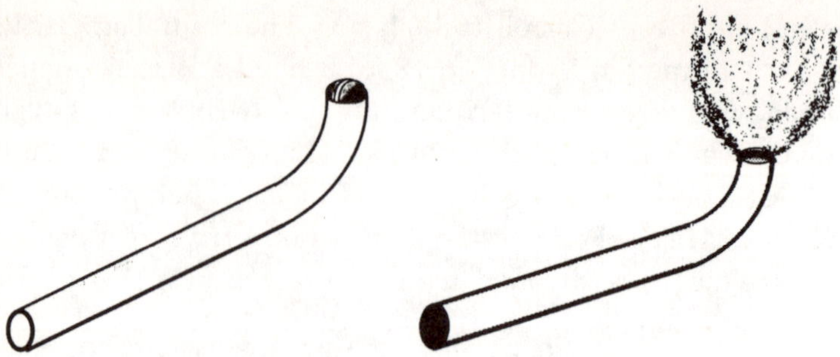

**Figure 45** The 'fishtail' gas burner has a slot in the end part of a gas pipe. The shape of the flame is due to the better supply of air at the ends of the slot. A few fishtail burners are still functioning. The Great Hall of the University of Sydney has some that are occasionally lit for 'atmosphere' on social occasions; however, the hall would be very ill-lit by the gas burners alone.

(Drawing by Peter R. Smith)

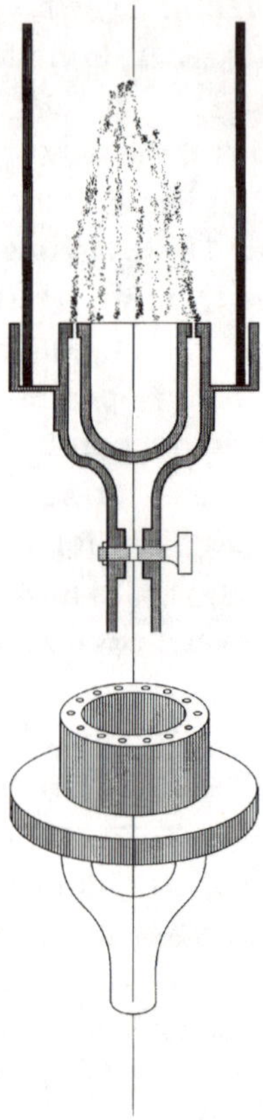

**Figure 46** Gas Argand burner. The gas supply is through the circle of holes; additional air is drawn in through the central hole.

(Redrawn by Peter R. Smith from Shar Jones [1984, p. 9])

However, gas lighting was widely used in Sydney for about forty years, and the light fittings gradually improved. In 1880 Sydney Town Hall installed a sixty-burner crystal 'gasolier', which was later converted to electricity.

In 1855 the German chemist Robert Bunsen invented the Bunsen burner, in which a predetermined quantity of air was mixed with the gas before it was ignited. This produced a hot and non-luminous flame. In 1885 another German chemist, Carl Auer von Welsbach, devised a mantle of silk fabric impregnated with thorium oxide, which turned incandescent when placed over a Bunsen burner. The Welsbach mantle was imported into Australia in 1887, and it transformed gaslight from relatively dim to brilliant illumination. By that time electric light had become available, although it was not yet widely used.

It took a few years before gas was adopted in the other capital cities: in Melbourne in 1850, in Hobart in 1854, in Adelaide in 1861, in Brisbane in 1864, and in Perth in 1885. Other towns followed, particularly those near the Victorian goldfields, and more than a hundred had built gasworks by 1913.

Although gaslight was eventually superseded by electric light, gas remained an economical source of energy for heating and cooking. A number of gas companies decided to eliminate competition from electricity by entering the field. The South Australian Gas Company secured permission to generate and supply electricity in 1882, and the Perth Gas and Coke Company in 1894 (Carroll 1988, p. 46).

## Electric light

The word electric is derived from *elektron*, an alloy of about four parts of gold and one part of silver, widely used by the ancient Greeks because they had difficulty in separating gold from silver. When the Roman Empire extended north, traders brought amber from the Baltic, and the Romans also called this *electrum* because of its similar yellow colour.

That amber, when rubbed with a suitable cloth, attracted or repelled some other substances was already known in Roman times, and during the Renaissance the name of amber was attached to this phenomenon. Thomas Browne, a seventeenth-century English physician, wrote 'by electrick bodies, I conceive . . . such as conveniently placed unto their objects attract all bodies palpable' (Oxford 1967, p. 591).

By the eighteenth century the phenomenon was generally called electricity, and intensively studied. However, there were few useful applications until the nineteenth century.

In 1810 Humphrey Davy demonstrated in a lecture at the Royal Institution in London that a bright light could be produced by an electric arc formed between two carbon rods.

One of the carbon rods that form the arc is slowly consumed, so that the distance between the rods needs to be adjusted from time to time. This was still done manually in the 1930s on some cinema and slide projectors; but for street lamps automatic adjustment was desirable. Patents were registered from 1845 onwards for clockwork mechanisms to keep the length of the arc constant by moving one of the carbon rods, and these were later improved by utilising the electric current.

In Davy's experiment the electricity for the arc light came from a battery, and until an electric generator was developed electric light had only a limited useful application. The first simple generator was produced in 1832, and by the 1850s generators were manufactured commercially.

The first practical use of a carbon arc light was in 1853 for the British South Foreland Lighthouse. It was employed in the 1870s for public lighting in France (for example, at the Gare du Nord and the Place de l'Opéra in Paris), and for football matches in England. Its first Australian use was in 1863 (see p. 119).

Arc lights remained in use for outdoor lighting until they were superseded by electric discharge lamps; but they were considered too bright for normal interior lighting. The incandescent light bulb was developed independently by Joseph Swan in England and by Thomas Edison in America in 1879.

Both consisted of glass bulbs from which the air had been evacuated, containing a carbonised thread that was rendered incandescent by the electric current, but could not burn because of the absence of oxygen (Figure 47). Both Swan and Edison contemplated legal action for patent infringement, but decided instead to co-operate.

The first electric power stations in England and in America were built to supply electricity for the new lamps, at Pearl Street in New York City, and at the Holborn Viaduct in London, both in 1882.

The first incandescent electric light bulbs gave no more illumination than the better gas fittings, such as the Argand burner (Figure 46). When the Welsbach mantle became available in 1885, gas actually provided brighter illumination; but electric light gradually gained in popu-

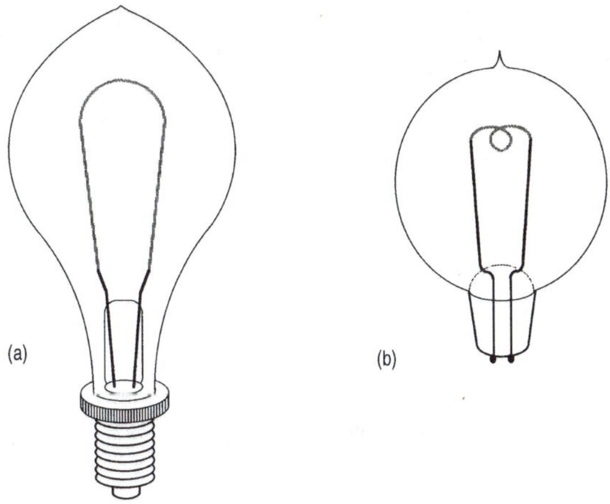


**Figure 47** The incandescent electric lamps invented (a) by Thomas Edison, and (b) by Joseph Swan, both using carbonised thread in an evacuated glass tube.
(Drawing by Peter R. Smith)

larity, because it was easily lit and did not produce any fumes, as Edison pointed out in a poster distributed with his light bulbs in 1882:

> This room is equipped with Edison Electric Light. Do not attempt to light with match. Simply turn key on wall by the door. The use of electricity for lighting is in no way harmful to health, nor does it affect the soundness of sleep. (Wells 1975, pp. 97–9)

The carbon filament was replaced by thorium oxide, then by tantalum, and in 1908 by tungsten. Each increase in the melting point of the filament permitted an increase in the operating temperature, and thus an increase in brightness. However, at these high temperatures the filament evaporated in a vacuum and blackened the inside of the bulb. From 1913 the bulbs were filled with an inert gas, at first nitrogen, and then argon. The filament was later coiled to reduce the loss of heat by convection in the gas-filled bulb, and in 1934 the coiled wire was coiled on itself to reduce the convection loss further (Figure 48).

Another type of electric lamp is based on an electric discharge. It had already been observed in the early eighteenth century that a glow could be produced by an electric discharge in a rarefied gas in a glass tube, and from time to time demonstrations were given before meetings of amateur scientists. In 1910 Georges Claude in France produced a neon-discharge tube, which gave a red light, and created a new industry

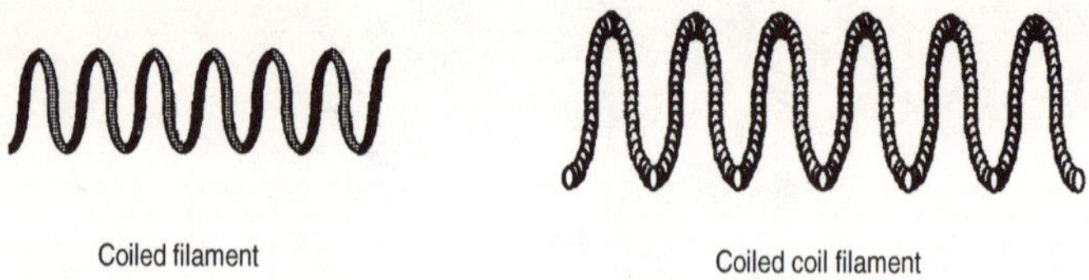


**Figure 48** The coiled tungsten filament, and the coiled-coil tungsten filament for electric light bulbs. (Drawing by Peter R. Smith)

of advertising in light. In the 1930s the very bright yellow sodium-vapour lamps became available, and they eventually replaced carbon-arc lamps.

Fluorescent tubes are mercury-vapour tubes with a fluorescent coating. Ultraviolet radiation from the conducting mercury vapour excites the fluorescent coating, which radiates light in the visible spectrum (Figure 49). The tubes were first demonstrated at the New York World Fair in 1938, but did not become commercially available in America until 1940, and in the rest of the world until after the war. They produce more light than incandescent lamps for the same amount of electric power. They are thus cheaper to operate, and produce less unwanted heat, which often has to be removed by air conditioning.

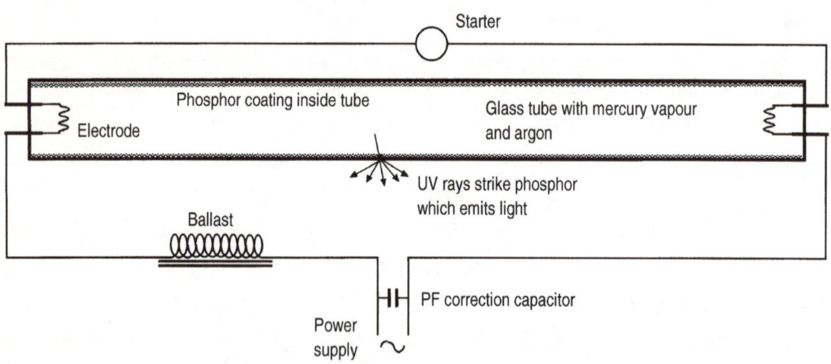


**Figure 49** The original fluorescent tube. The lamp, shown on the top of the figure, consists of a glass tube coated inside with a phosphor that emits visible light when struck by invisible ultraviolet rays. These are produced by an electric discharge between the two electrodes, set up in the column of mercury vapour and argon gas inside the tube. The lamp needs a starter and a heating circuit for the two electrodes.

(Drawing by Peter R. Smith).

## Electricity in Australia

Gas has only a few useful applications, of which lighting was initially the most important. Electricity, however, soon proved versatile. For Australia its most significant early use was in telegraphy. The electric telegraph service started in Melbourne in 1854 (see p. 184), and by the mid-1860s the major post offices had electric batteries powerful enough to operate the service (Atkinson 1875, pp. 749–766).

The first railway was built from Melbourne to Port Melbourne, also in 1854. Railways were soon built in every state, and they required some electric services. The first substantial demand for electric power came, however, with the electrification of the trams.

The first trams in Sydney in 1861 were horse-drawn. Steam trams were introduced, also in Sydney, in 1879, and steam-operated cable trams in Melbourne in 1885. An experimental electric tram was built in 1889 in Box Hill, on the outskirts of Melbourne; it operated until 1896 (Edwards 1969, p. 13). Another operated between Randwick and Waverley in Sydney from 1890 to 1892. Between 1893 and 1899 several electric tram services were built on Sydney's North Shore as extensions to existing cable trams (Fraser 1989, p. 265).

In 1897 it was decided to electrify the entire Sydney tram system, and a central power station was built at Ultimo (Figure 50), which began operation in 1899. The Melbourne electric tram system started operation in 1903.

By 1903 electricity was therefore put to substantial use in Australia. However, neither Sydney nor Melbourne had established a city-wide system for electric lighting.

## Electric lighting in Australia

The first electric lights in Australia were carbon-arc lamps set up at the Electric Telegraph Office and on Observatory Hill, Sydney, on 11 June 1863, when the city was otherwise illuminated with gas and oil lamps in honour of the wedding of the Prince of Wales (Edwards 1969, p. 10). The power for the electric lights was supplied by batteries.

In 1878 electric-arc lighting was used to permit after-dark work during the construction of the 'Garden Palace' (see p. 56) at the request of the builder, John Young. By that time an electric generator was available.

**Figure 50** The Ultimo Power Station operated from 1899 to 1963. In the 1970s it was decided to hand the building over to the Museum of Applied Science to establish the Powerhouse Museum; this photograph was taken after that decision, but before building commenced. Some of the old power station's ironwork and steelwork, including the hoisting gear, have been retained inside the new museum. (Photograph reproduced by courtesy of the Powerhouse Museum)

The new Macquarie Lighthouse, replacing Greenway's 1817 structure, was opened in 1883 (see Figure 18). The old lighthouse had been lit by oil. The new one had an electric-arc light; the electric generator was powered by a gas engine, and a special gas holder was built to regulate the pressure of the gas supplied by the Australian Gas Light Company (see p. 113).

In 1888 the Victorian Government organised a celebration on the centenary of Australia's first settlement in the Exhibition Hall, built eight years earlier for the Melbourne International Exhibition. On this occasion the building was lit by a thousand electric arc lamps (Edwards 1969, p. 12).

All these special uses of electric lighting had been organised by the government. However, during the late nineteenth century electric lighting depended mainly on private enterprise. Electric power, unlike gas, could be produced on a small scale, using relatively small generators powered by a steam engine or by gas. Both in Sydney and in Melbourne electric power for lighting was therefore initially generated by a number of small companies. The first in Melbourne was the Victorian Electric Light Company, which established a small power station in Russell Place, off Bourke Street, to supply light for the Eastern Market and the Athenaeum, and later to some additional domestic and commercial customers (Leeper 1955, p. 154). In Sydney also there were several small private electricity companies.

In 1888 Tamworth (NSW) became the first Australian town to use electricity for general street lighting. This was only six years after London and New York. The Tamworth service was limited to street lighting. There was already a gas company which retained the right to supply domestic and commercial customers until 1904, when electricity became generally available in the town (Brady 1994, p. 6).

In 1889 Young (NSW) established the first public electricity supply in Australia, which served offices and homes in addition to providing street lighting (Brady 1996).

In 1896 the Municipal Council of Sydney Light Bill was passed, with the object of providing street lighting with electric-arc lamps, and interior lighting for private customers. However, political disputation over whether the power system should be developed by private enterprise or by the Municipal Council caused considerable deferments. Pyrmont Power Station was not built until 1904. The electric cables under the streets were laid by a German company, Kabelwerk Duisburg, and the street lighting was switched on by the Lady Mayoress in 1904. Pyrmont Power Station was the first major power station to provide a public electricity supply; the main function of Ultimo Power Station (see Figure 50), built in 1899, was the supply of power for the trams.

In Melbourne also there was a dispute about whether the electricity should be supplied by a private company or a public authority. The town supply was originally provided by the private Nhill Electric Light Company in 1891; in 1899 it was taken over by the Melbourne City Council (Edwards 1969, p. 14).

Victoria lacked a supply of black coal, and it imported it from New South Wales. This added to the cost, and it rendered it susceptible to delay by frequent strikes over which it could exercise no control (Edwards 1969, p. 15). In 1911 two experts were invited to advise on the potential use of the extensive brown-coal deposits in the Latrobe Valley; Merz was an electrical engineer from England, and Hoffmann a mining engineer from Germany, which had made extensive use of brown coal for several decades. In 1918 the State Electricity Commission was established to utilise the brown coal for electricity generation, and in 1920 John Monash (see p. 96) became its general manager. By 1960 approximately 85 per cent of Victoria's electricity was generated from brown coal in the Latrobe Valley.

The advantages of larger units for electricity generation were evident also in New South Wales. The Sydney County Council was formed in 1935. In 1950 the New South Wales State Electricity Commission was created. It located the new power stations near the coal, instead of transporting the coal to the population centres. Thus the inevitable pollution from the combustion of coal occurred in less sensitive locations, and electricity could be transported more cheaply than coal.

# 6

# THE THERMAL ENVIRONMENT

## Different energy problems of heating and cooling

Buildings have been heated since the early days of civilization. This can require a great deal of energy, far more than lighting. A hundred-watt bulb gives a very bright light; but a thousand-watt electric radiator is only a small heater.

Air conditioning was in the 1920s and 30s even more lavish in the use of energy, because the air has to be cooled *below* the desired temperature to reduce the humidity, and it must then be reheated to that temperature. However, air conditioning did not become widely used in Australia until the 1960s, and even the first American air-conditioning installations were built only in the 1920s. Before then hot-weather comfort, in so far as it could be achieved, depended mainly on constructional devices that did not involve the use of energy.

The significance of the energy consumed by buildings became evident in 1973 when, following an unsuccessful attack on Israel by its Arab neighbours, the Middle Eastern oil-producing countries restricted the supply of oil and trebled its price. This resulted in a determined move by the oil-importing countries to reduce their consumption by energy conservation and by the use of alternative energy sources (see p. 144).

The resulting investigations (Cowan 1980, p. 4) showed that in Britain about 50 per cent of the total energy was consumed by buildings, the major part for heating. In America the absolute energy consumption per person for buildings was even higher, but the proportion (40 per cent) was lower because of the much greater use of motor cars. In Australia buildings accounted for only 20 per cent of the national energy consumption, even though there was by 1973 a significant use of air conditioning. This relatively low figure was due to the mild climate of the densely inhabited parts of Australia.

Captain Watkin Tench, who arrived with the First Fleet, commented on that already in 1791 (see p.17). In his day the European settlement was confined to a coastal strip in the subtropical zone; but this is still largely true today, even though the densely inhabited strip is now wider and extends much further around the coast. There are many parts of Australia that have an extreme climate, but as few people live there they do not exert a significant influence on the energy consumption in buildings.

The position is different in America and in Europe, where large cities, such as Chicago, experience both very hot and very cold weather each year, so that a great deal of heating is essential, and air conditioning becomes very desirable.

In many of the densely inhabited parts of Australia, on the other hand, it is possible to be comfortable with only a little winter heating and without air conditioning by designing buildings appropriately. These measures rely indirectly on the energy of the sun, and they are therefore today often called 'passive solar energy'.

## The classical tradition

The ancient world discovered empirically some of these rules. Socrates' views, expressed in the fourth century BC in a conversation with Aristippus, were recorded by Xenophon:

> Again his dictum about houses, that the same house is both beautiful and useful, was a lesson in the art of building houses as they ought to be. He approached the problem thus:
>
> 'When one means to have the right sort of house, must he contrive to make it as pleasant to live in and as useful as can be?'
>
> And this being admitted, 'Is it pleasant,' he asked, 'to have it cool in summer and warm in winter?'
>
> And when they agreed with that also: 'Now in houses with a south aspect [in the northern hemisphere] the sun's rays penetrate into porticoes in winter, but in summer the path of the sun is right over our heads and above the roof, so that there is shade.

'If then this is the best arrangement, we should build the south side loftier to get the winter sun, and the north side lower to keep out the cold winds. To put it shortly, the house in which the owner can find a pleasant retreat at all seasons and can store his belongings safely is presumably at once the pleasantest and the most beautiful.' (Xenophon 1968, pp. 22–3)

Vitruvius, writing about three hundred years later for a more affluent readership, recommended that individual rooms should be oriented specifically to fit their purpose and the season:

> Winter dining rooms and bathrooms should have a southwestern exposure for the reason that they need the evening light, and also because the setting sun, facing them with all its splendour, but with abated heat, lends gentler warmth to that quarter in the evening. Bedrooms and libraries ought to have an eastern exposure, because their purpose requires morning light, and also because books in such libraries will not decay . . .
>
> Dining rooms for spring and autumn to the east; for when the windows face that quarter, the sun, as he goes on his career from over against them to the west, leaves such rooms at the proper temperature at the time when it is customary to use them.
>
> Summer dining rooms to the north, because that quarter is not, like the others, burning with heat during the solstice, for the reason that it always keeps cool, and makes the use of the rooms both healthy and agreeable. (Vitruvius 1960, pp. 180–1.)

Vitruvius, of course, was writing for the northern hemisphere, and for readers who could afford a house with three rooms suitable as dining rooms. With that proviso, this is still good advice for a house without air conditioning.

Similar rules were published by various writers from the Renaissance on, and they were utilised in English eighteenth-century architecture, which was one of the early influences on Australian building construction.

However, another early British influence was the importance attached to the fireplace as an essential feature of a living room.

## Heating

The simplest way of heating a building is to put a fire into the middle of the room, and allow the smoke to escape through a hole in the roof, which may be covered by a cowl. This method was still used when the Great Hall of Hampton Court Palace was built in England in the 1530s; the location of the fireplace can still be seen, although the hole in the roof has been closed.

The chimney was probably invented in the thirteenth century in fortified castles, where the roof of the keep and the room immediately below it were needed for defensive weaponry. Since the living quarters were thus not immediately below the roof, the smoke could not be exhausted through a hole in the roof, and the fireplace was built into the wall, with a chimney that allowed smoke to escape through the outer wall. This had obvious advantages, and the fireplace with a chimney became the most common method of heating in Britain (Gilbert & Wells-Cole 1985, p. 5), and hence in Australia. Virtually all the public buildings and the houses of the officers and persons of means in the early days of the colony had at least one fireplace (see p. 18).

In the nineteenth century English fireplaces generally burned coal, which was a major cause of air pollution. In Australia the fuel was commonly wood. The cheerful appearance of the open fire and the smell of wood may have contributed to the continued popularity of fireplaces, which were still the most common form of heating in Australia in the early twentieth century, in spite of the amount of work involved in cleaning them and laying the next fire. In the wealthier houses the fireplaces often had mantelpieces of polished marble or elaborately carved cedar.

Portable heaters ultimately proved more convenient for domestic heating in the relatively mild winters of many Australian cities. They were easily turned on and shut down, and they could be moved from room to room, whereas the fireplace was confined to one room. Kerosene heaters were gradually replaced by gas heaters and by electric radiators, and then by electric convection heaters on wheels, which posed less risk for children.

Two other methods of heating eventually became established in Australia. The hot-water radiator was originally developed in the early nineteenth century. James Watt (see p. 59) was one of the first to install one in his office, and they became common for heating larger buildings, such as hospitals and schools. By the end of the century hot-water radiators came into use in Australia (Figure 51) for hospitals and other public buildings, and for multi-storey apartment and office buildings. They are still used today, as are hot-water coils as a source of heat for the larger air-conditioning installations.

**Figure 51** Cast-iron hot-water radiator. In the nineteenth and early twentieth centuries radiators were often quite ornate. The hot water for all the radiators was supplied by a central boiler.
(Reproduced from an advertisement in Haddon [1908, p. xxii])

In America, where many cities have prolonged spells of very cold weather, requiring the heating of the entire house, the plenum duct was developed in the late nineteenth century. This is at a slightly higher pressure than atmospheric, and through grilles supplies hot air from a central furnace. Plenum ducts, however, were not used in Australia until air conditioning was introduced.

Heat pumps were originally proposed by William Thomson (later Lord Kelvin) in 1852 (Billington & Roberts 1982, p. 275) to be used for heating or for cooling by reversing the cycle (Figure 52). By the time air conditioning became common in Australia, they had become a practical possibility (see p. 139) and they eventually replaced the traditional methods of space heating in many of the smaller air-conditioned buildings.

In Britain the hot-water supply was generally an adjunct to the general heating system. The most common method in nineteenth-century houses was to place a boiler in the living-room fireplace, which produced a hot-water supply as long as the fire was lit. This was adequate for the kitchen, for washing, and for a bath once a week for each person in the house. This system was also used in Australia, but with the boiler in the kitchen range (Figure 53)

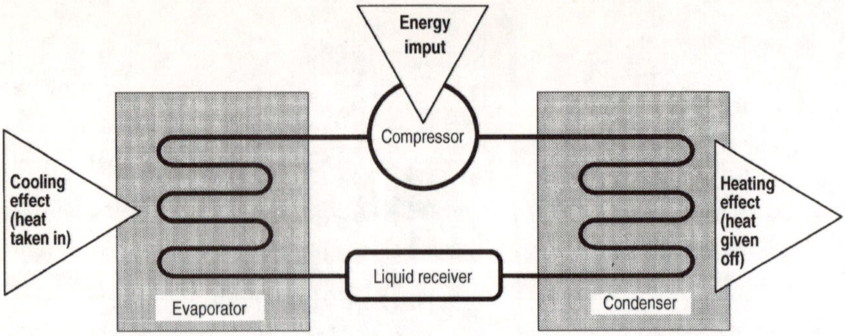


**Figure 52** The principle of the heat pump: The heat pump can be used for cooling the air or, in reverse cycle, for heating, when the heat rejected becomes the useful output. Although developed in America, the first reverse-cycle installation was built in Australia in the late 1950s for the Kogarah (NSW) Telephone Exchange (Carrier 1982).

(Drawing by Peter R. Smith)

Water heating is relatively more important in Australia, where space heating is only needed for a few months, while the daily bath or shower became an established custom relatively early. Separate water heaters have therefore become customary.

The earliest separate water heaters were fixed near the bath, and the water was heated by wood chips burned in a metal fire box; these heaters were still installed in the late 1940s. However, gas generally replaced wood as a fuel, where and when it became available. A central water heater, which supplied the kitchen, the bath and the wash basins, was already used in the late nineteenth century (Figure 54).

In most of Australia the hot-water supply uses more energy than space heating, and fuel economies can be achieved by the use of solar energy (see p. 145).

## Thermal comfort in warm weather

Comfort in cold weather is easily achieved by heating, or by putting on warmer clothes, or by a combination of both. However, a mechanical lowering of the air temperature did not become common until the twentieth century, and during the nineteenth century gentlemen were expected to dress formally in public, and often also at home; respectable women wore long skirts and sleeves. Comfort in warm weather therefore presented a much greater problem.

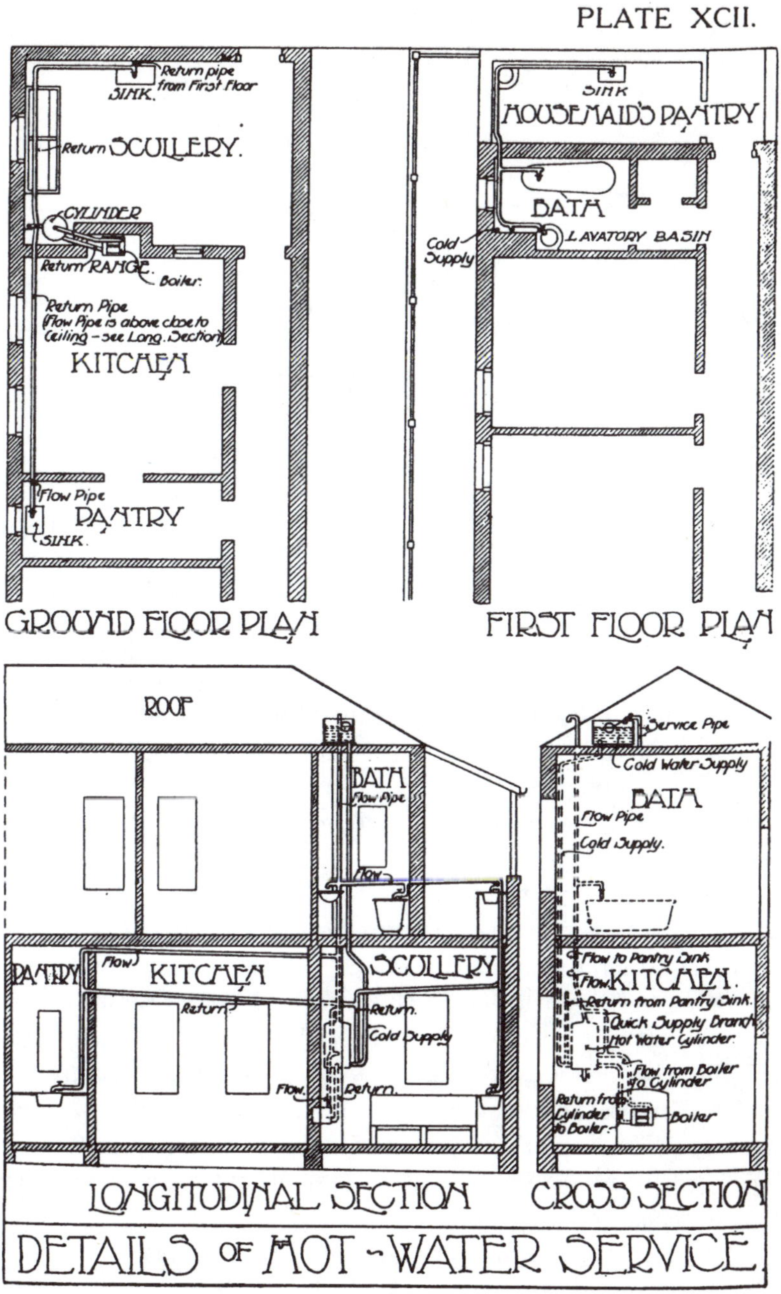


**Figure 53** Hot-water service illustrated in one of the few textbooks on architecture (Haddon 1908, p. 461) published in Australia before World War I. It is the only domestic hot-water system discussed by Haddon.

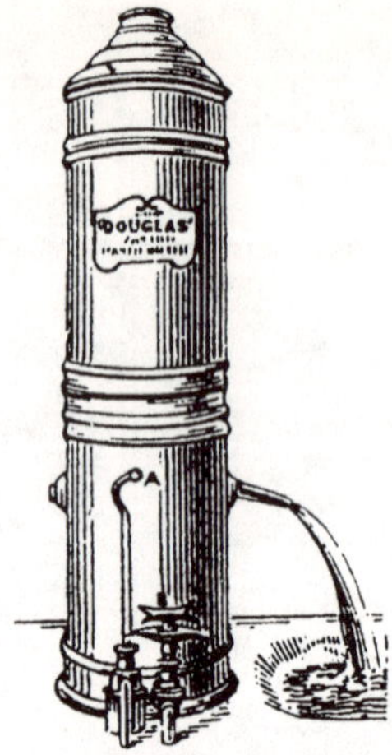


**Figure 54** Separate water heater, illustrated in an advertisement (Haddon 1908, p. xx). In 1908 the manufacturers offered heaters using wood, kerosene, or gas. Electric water heaters became available later.

The temperature of the central part of the human body is approximately 37°C, and in conditions of thermal comfort the skin temperature is approximately 33°C. The energy produced in the body by the oxidation of food is converted into heat, except for that used for performing work. This heat must be transferred to the surrounding air.

Sweat evaporates quickly in hot dry weather, and the latent heat of evaporation cools the skin. One can therefore be quite comfortable in a hot-arid climate at high temperatures if one drinks enough water.

Sweat does not evaporate so readily in high humidity, and one can be very uncomfortable at temperatures well below 37°C in still air. Even slight air movement causes heat transfer by convection from the skin and from the clothing, and this provides great relief in warm-humid weather. This has been known empirically since antiquity. Egyptian tomb paintings show servants with big fans standing behind their masters.

Thermal comfort in hot weather without artificial cooling is therefore promoted by

- preventing the entry of solar radiation by sun shading
- by improving the insulation of the walls and the roof of the building
- in warm-humid climates, ventilation to provide air movement and/or
- in hot-arid climates, evaporative cooling.

## Ventilation and the veranda

The climate of Sydney, and indeed of most of the densely inhabited regions of Australia, is warm-humid rather than hot-arid. The First Fleet arrived in mid-summer, and the convicts at work and the formally-dressed officers must have suffered from the warmth and humidity; some commented on it in their diaries (Clark 1981, p. 91).

It took time, however, to develop a response in the early buildings, which had little sunshading; and the first veranda was not built until late 1792 or early 1793 (see p. 17). The advantages of the air movement on the open veranda on humid summer days must have been immediately obvious, and during the early nineteenth century a veranda became an essential part of the better Australian houses.

Verandas were already well established in British colonial architecture; indeed, the word is taken over from Hindi. In India the only thermal comfort problem was caused by hot weather. If the veranda was very wide and ran around the house, as it often did, it kept it in shade all the year round. It was also wide enough to provide a proper open-air living space. One could eat, sleep, entertain, and even study or work on a wide veranda during the hot summer period.

However, in New South Wales the hot weather lasts only a few weeks, and there is a similar period when the heat generated in the interior of the house by penetration of solar radiation is pleasant.

The wide veranda makes the interior of the house in winter unnecessarily cold and dark because of excessive shading, and it is not the best solution for year-round thermal comfort. The ancient Greeks and Romans, who lived in a climate not unlike that of Sydney, had made due allowance for sunlight penetration in winter (see p. 125) to allow for the utilisation of the 'passive' solar energy, and this concept was revived during the Renaissance. However, it was not utilised in Australia until the later part of the nineteenth century. Before then most verandas were made too wide to permit sunlight penetration into the house in winter.

The wide veranda was very popular for summer living for several decades. Paintings, and later photographs, several reproduced by Philip Drew (1992), show the extent to which individuals, and even complete families made use of the veranda for every kind of activity, including sleeping. Drew described the veranda as 'a touchstone of Australia's national identity, a wellspring of memories, and a bridge linking the colonial past with the future'.

However, during the twentieth century verandas have been used less for outdoor living. They continue as favourite places for recreation on a hot day or for a summer-evening party, but dining or sleeping on the veranda has become less common. Verandas became narrower to allow more daylight into the interior, and some houses were specifically designed to permit sunlight penetration to warm living rooms in winter, at first empirically, and after World War II by calculation, graphic construction, model analysis, or later by a computer program (Cowan & Smith 1983, pp. 41–52).

The change was partly due to a more informal style of living. People no longer wore heavy clothes to the same extent at home, and could therefore be comfortable with the lower air velocities that were attainable indoors. Another factor was the advent of insect screens. These could also be used to screen a veranda, but it then lacked the open communication with the exterior (Figure 55).

There was consequently a new emphasis on providing through-ventilation for the interior of the house by an appropriate alignment of doors and windows, as people chose to live in a well-ventilated and insect-screened house rather than on an open veranda during the warm-humid weather.

**Figure 55** Veranda of Vaucluse House, Vaucluse, NSW, built by William Charles Wentworth in the 1830s, a living space between the house and the garden.
(Photograph by Henry J. Cowan)

Good ventilation depended on providing both an entry and an exit for the breeze, and arranging the opening to catch the prevailing wind (Figure 56). Along most of the east coast of Australia this comes from the north-east on hot summer afternoons; in Perth the 'Fremantle Doctor' is a south-westerly breeze.

Artificial ventilation played only a minor part in the design of Australian buildings until the late nineteenth century, in spite of its use in India and in Britain. In India the punkah, a slow-moving, large cloth-covered frame, was hand-operated. In British factories fans were driven, together with other machinery, by steam engines from the early nineteenth century on. By the mid-nineteenth century fans operated by steam engines specially installed for the purpose were employed for the artificial ventilation of hospitals through air ducts (Cowan 1978, p. 240). By the end of the nineteenth century a number of hospitals in industrial cities in Europe and America employed air filters with artificial ventilation and air heating. This was, in fact, air conditioning, but without air cooling.

A search in medical libraries and questions to medical practitioners have produced no evidence that ducted ventilation or air cleaning was ever used in Australian hospitals until full air conditioning was introduced in the 1930s. Hospitals relied until then for their ventilation mainly on open windows and on verandas. Some still prefer natural ventilation at the present time, on the ground that air-conditioning ducts are liable to harbour bacteria.

Electric fans, however, were adopted when they became available (Figure 57).

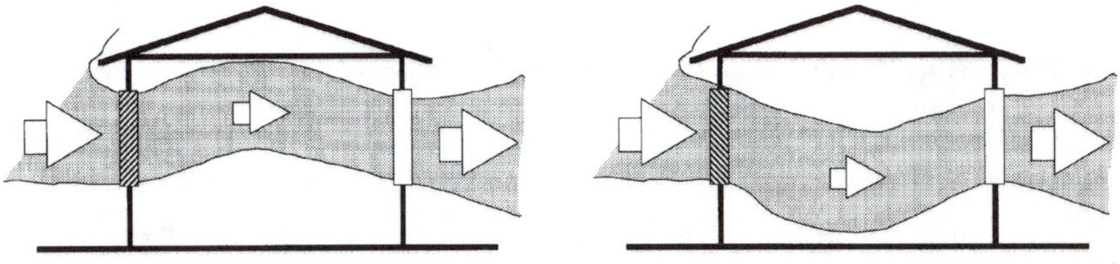

**Figure 56** Natural ventilation requires both an entry and an exit to be effective. This can be achieved by placing windows on opposite sides of the same room, or by suitably oriented internal and external doors. The direction of the air flow can be influenced by the angle of the windows or louvres through which the breeze enters.

(Drawing by Peter R. Smith)

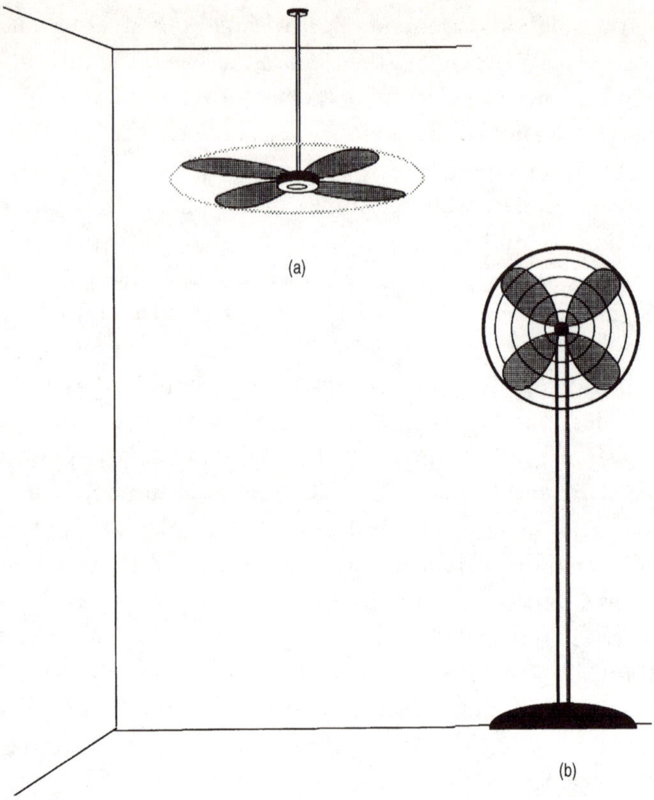


**Figure 57**

(a) Slowly rotating ceiling fan: This type has the same advantages as the hand-operated Indian punkah—the fan noise is low, and it does not cause papers to fly around.

(b) Portable fan: The man-sized type can produce air velocities up to 4 m/sec (a 'light breeze' on the Beaufort scale) at 3 metres from the fan, and 0.5 m/sec at distances up to 12 metres. The desktop fan is more mobile, but has a smaller range.

(Drawings by Peter R. Smith)

## Sunshading, insulation, and evaporative cooling

In southern Australia an eaves overhang, a balcony above, or some other projection of the order of one metre (depending on the latitude) on a north-facing wall provides sunshading in summer, while allowing sunlight penetration for winter heating. By the mid-nineteenth century a number of buildings seem to have been deliberately designed to take advantage of this form of solar energy. The exact amount of the projection needed in any particular location would have become known to local builders and architects from experience.

A number of homesteads and city mansions, such as Elizabeth Bay House, had master bedrooms and 'morning rooms' facing approximately east. Entry of the western sun was generally considered a disadvantage, but it could not be prevented by a horizontal projection (Figure 58). Some buildings had blank west walls, or small west-facing windows.

After World War II theoretical design methods for sun shading came into use, and so did vertical sun louvres to control the western sun. The energy crisis of 1973 (see p. 123) further encouraged their utilisation.

During the nineteenth century deciduous trees were commonly planted (Figure 59) that gave good shade in summer while allowing sunlight penetration in winter. However, after World War II the planting of deciduous trees was discouraged on the ground that they were foreign to the Australian landscape. This had an adverse effect on the interior thermal environment of the buildings.

Thermal insulation played a limited role as an aid to the thermal environment before 1973. Watson Sharp (1969), the author of a widely-used book on Australian building construction mentioned only cork, softboard and plywood in 1969, even though by the 1960s lightweight insulating materials, such as glasswool and various plastic foams, had become available. They had very low thermal conductivities, because they contained pockets of trapped air (Cowan & Smith 1983, p. 76).

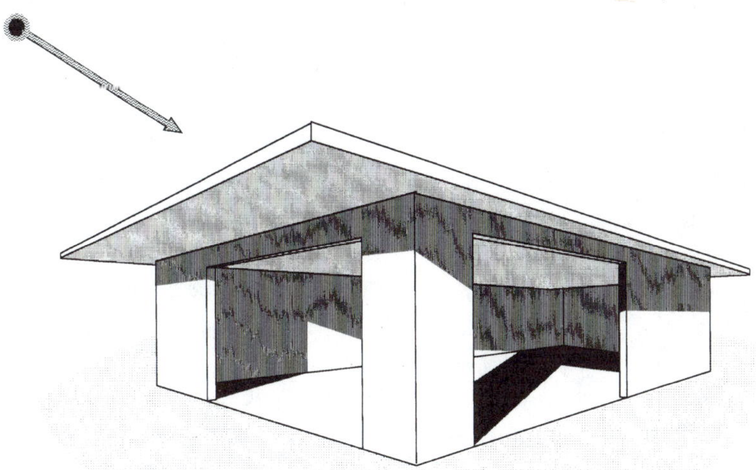

**Figure 58** Horizontal sunshades are not effective against the western sun, because its altitude is too low in the west. The solar radiation is approximately the same from the eastern morning sun, but in the early morning it is pleasant, whereas in the heat of a late summer afternoon it may be unwelcome. (Drawing by Peter R. Smith)

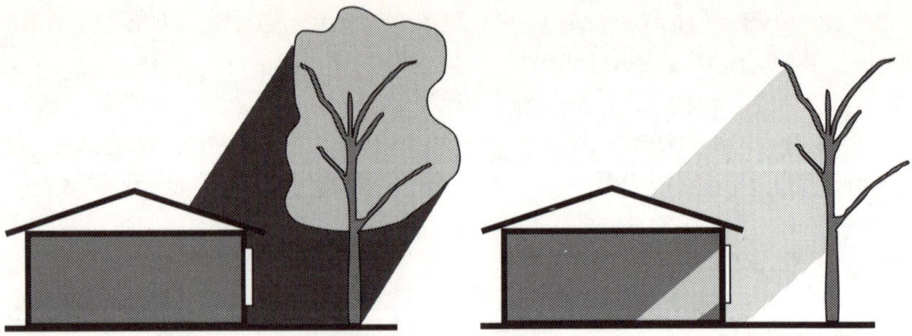

**Figure 59** Deciduous trees can provide shade in summer, while permitting sunlight penetration in winter.
(Drawing by Peter R. Smith)

These were the materials mainly used in North America and in Europe after 1973 to improve the thermal insulation of buildings to reduce the energy consumption of buildings, and thus reduce fuel imports. Some overseas authorities imposed rules that required minimum insulation standards for new buildings, and retrofitting of old buildings. Others used tax incentives to encourage owners to undertake the work without compulsion.

These measures were widely reported in Australia, and standards of thermal insulation (to reduce heat transfer by conduction) greatly improved. In Australia heat transfer due to solar radiation is also important, and this can be reduced by *reflective insulation.* Bright aluminium foil was introduced in the 1960s to act as reflectors (Hassall 1969, pp. 22–4).

In the hot-arid regions *evaporative cooling* (Figure 60) was employed to some extent to reduce the temperature and increase the humidity before air conditioning was introduced, and it is still used in buildings, such as schools, for which air conditioning is considered too expensive.

Ever since antiquity *pools of water and fountains* have been built in warm-weather countries. They became focal points of urban design in many European cities after the Renaissance.

While the temperature reduction they cause in the open air in a humid climate is negligible, they convey a pleasant impression of coolness, and it is a matter for regret that so few have been built in Australian cities. The backyard swimming pools constructed in private houses since the 1960s have become our substitute, the Australian version of the fountains of Rome.

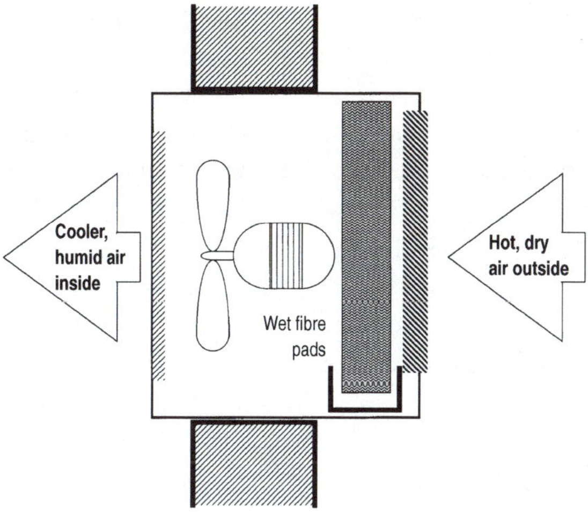


**Figure 60** Evaporative cooler, consisting of water-soaked fibrous pads placed on the suction side of a wall-mounted fan. It is useful in hot-arid regions.
(Drawing by Peter R. Smith)

## Refrigeration and air conditioning

Direct cooling was already practised in antiquity, but it was expensive. The most common method was to collect ice or snow from the surrounding countryside in winter, or from mountains at any time, and store it in a deep hole in the ground, covered by an insulating layer, such as straw. The Greeks developed a freezing mixture, and the Egyptians, who had no access to ice or snow, discovered how to cool drinks by evaporative cooling (Billington & Roberts 1982, p. 239).

Evaporative cooling was still used for food preservation in Australia in the twentieth century in locations where refrigerators and ice were not available. The 'Coolgardie Safe' is a wire cage covered with a damp cloth; as the water evaporates the temperature in the 'safe' is lowered.

In the late Middle Ages ice harvesting became a regular winter practice in European countries north of the Alps, and deep cellars were built to store it. By the eighteenth century many French and English country estates included 'ice houses', often decorative pavilions above ground, which stored ice blocks in an underground cellar.

In the late eighteenth century an American shipowner started exporting ice from the northern to the southern states. This eventually developed into a worldwide trade. In the mid-nineteenth century large quantities of natural ice were exported from the American east coast, some around Cape Horn into the Pacific, and Sydney and Melbourne were among the destinations. The harvesting was mostly done by lumber contractors as an off-season activity. They had the saws, the carts, the draught animals, and the sawdust needed for insulation.

Natural ice was used on the new American railways to transport refrigerated foodstuffs, and a little later for cooling passenger cars. It was used to refrigerate food on transatlantic passenger ships, but it all melted before the completion of the much longer voyages through the tropics from Australia, New Zealand or Argentina to Europe. A great effort was therefore made to develop refrigerating machines, so that the meat export market could be developed.

Refrigeration machines used the latent heat required to evaporate a liquid for cooling; this latent heat was absorbed from the cooling medium. All machines in use were closed systems, that is, the gas was condensed to a liquid and re-used. The machine therefore had an evaporator, a condenser, and a source of power (Figure 61). This is now usually an electric motor, but in the eighteenth century the work-input

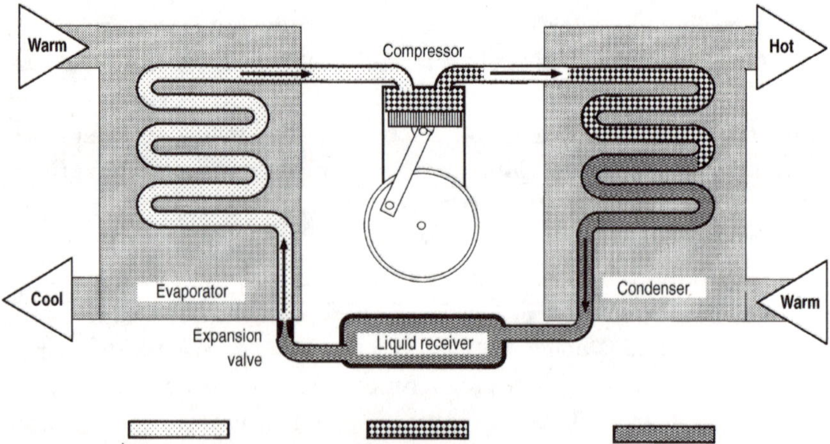


**Figure 61** Principle of the mechanical, or vapour-compression, refrigeration cycle. The refrigerant is kept under pressure in the liquid receiver. It is allowed to expand through a valve into the evaporator, which is at a lower pressure. At this lower pressure it vaporises and takes on latent heat. This cools the evaporator. A compressor then compresses the vapour. This raises its temperature, but it does not liquefy until it is cooled in the condenser. Heat produced in the condenser is available for use, if needed; otherwise it is discarded.

(Drawing by Peter R. Smith)

usually came from a hand-pump, and in the nineteenth century it was commonly supplied by a steam engine.

James Harrison is widely acknowledged as the pioneer of commercial refrigeration (Carroll 1988, p. 59). He made his first artificial ice to preserve fish that he caught commercially in Port Phillip Bay, using a metal coil, ether and a hand-pump. From that he developed a steam-operated refrigeration machine, which started to make ice for sale in Geelong in 1859 (Figure 62), the first to compete with the natural ice imported from America. He later designed refrigeration machines for shipboard use that eventually led to the successful export of Australian meat to Europe.

By the end of the nineteenth century smaller refrigeration machines became available. Early domestic refrigerators were powered by kerosene or gas (using the absorption refrigeration cycle, which has no significant moving parts), and only later by electricity (using the vapour-compression cycle). The first were sold in the USA in 1895, and the first in Australia in the 1920s, but ice-boxes for keeping food refrigerated were used until after World War II, although the ice was no longer natural, but machine-made.

Ice was also used to cool buildings, but this had only moderate success. A notable example was the air-cooling system installed in the

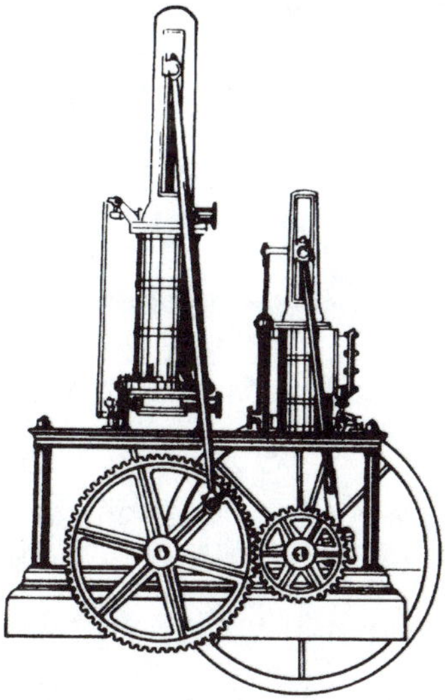

**Figure 62** Vapour compression machine made by James Harrison in Geelong (Vic.) in 1859 for making ice. It was powered by a steam engine.
(Reproduced from Woolrich [1969])

Carnegie Concert Hall, New York, when it was built in 1889 (Elliott 1992, p. 316). Air was drawn in by four steam-powered fans over racks of ice blocks 4 metres high. This lowered the air temperature appreciably, but it increased the humidity. As the summer weather in New York City is normally humid, the thermal comfort was not greatly improved.

By 1900 mechanical refrigeration had mostly superseded cooling with natural ice, because it was simpler and more effective, and it was used successfully for the textile, chemical and food-processing industries.

However, air conditioning for human comfort requires not merely refrigeration, but also dehumidification of the air. The problem is best understood by using a psychrometric chart (Figure 63), a graphical presentation devised by Willis Carrier, the principal inventor of air conditioning (Carrier 1965, p. 1.172). It shows that, as air is cooled without altering its moisture content (the *absolute* humidity), the *relative* humidity increases to unacceptable levels, and condensation may occur. In a

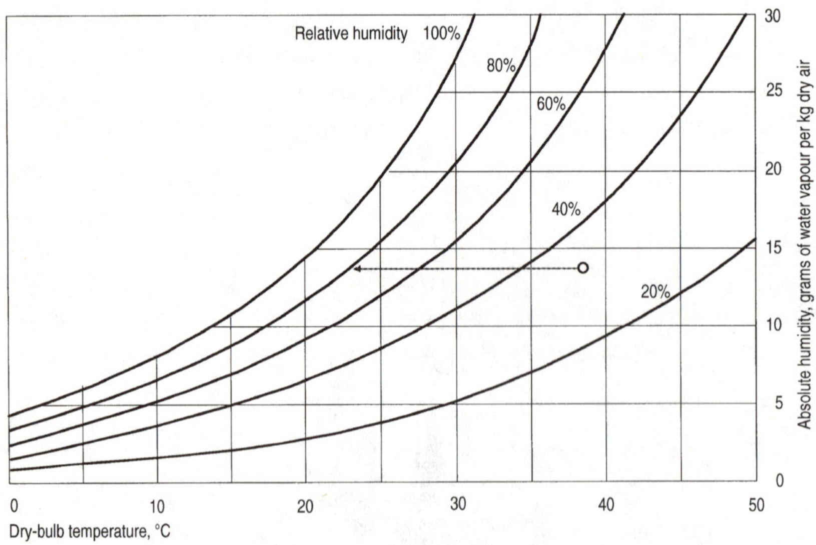


**Figure 63** Psychrometric chart; the word comes from the Greek *psychros* (chill). The dry-bulb temperature (the normal temperature) is plotted on the horizontal axis, and the absolute humidity (the moisture content) on the vertical axis. The relative humidities resulting from a change in temperature are shown by the curved lines. At a relative humidity of 100 per cent the water vapour condenses, and that is called the dew point.

The horizontal line terminating in an arrow shows that if air at a temperature of 38°C and a relative humidity of 35 per cent is cooled to 23°C without altering its moisture content, the relative humidity increases to 80 per cent. At 18°C the dew point is reached. If the air is cooled further, some of its moisture condenses.

(Drawing by Peter R. Smith)

refrigerator the condensed moisture forms a layer of ice, which in modern refrigerators is removed by automatic de-icing. In air to be used for a room for human occupation it must be removed beforehand.

There are a number of ways in which this can be done, but Carrier's dewpoint method is now commonly used. According to his own account, Carrier had the idea while observing steam at a railway station while waiting for a train:

> Here air is approximately 100 per cent saturated with moisture. The temperature is low so, even though saturated, there is not much moisture. There could not be, at so low a temperature. Now, if I can saturate the air and control the temperature at saturation, I can get air with any amount of moisture I want . . .
>
> When very moist air is desired, I'll heat the water. When very dry air is desired, that is, air with a small amount of moisture, I'll use cold water to get low-temperature saturation. The cold-water spray will actually become the condensing surface. I certainly will get rid of the rusting difficulties that occur when using steel coils for condensing vapour in air. Water won't rust. (Ingels 1952, p. 15)

Air conditioning therefore requires cooling the air to a lower temperature, the appropriate dew point, at which the excess water condenses, and then reheating it to the desired temperature.

Carrier was only one of a number of engineers working on air conditioning, but he was the most successful in designing installations that increased human comfort, as distinct from merely lowering the air temperature. Carrier's first fully air-conditioned auditorium was Graumann's Metropolitan Theater in Los Angeles in 1922. The first high-rise office block to be fully air conditioned, in 1927, was the 21-storey Milam Building in San Antonio in Texas. The conditioned air was supplied through ducts above the corridor ceilings, and the air return through grilles in the office doors; the corridors acted as air-return ducts.

By 1931 four hundred American theatres had been air conditioned; most were cinemas whose owners were more receptive to technical innovation than those of live theatres. Individual rooms in a number of buildings were air conditioned before World War II, but only a few complete buildings. Rockefeller Center in New York, started in 1931 and completed in 1939, was not fully air conditioned, but some parts of it were, notably the Radio City Music Hall. This remained the showpiece until the late 1940s. Le Corbusier (1937, p. 33) referred admiringly to its 'conditioned air throughout, pure, clean, and at a constant temperature' in *When the Cathedrals Were White*. It was the technical realisation of one of Le Corbusier's visions, but which owed nothing to him.

The first theatres to be air conditioned in Australia were the Prince Edward Theatre in Sydney in 1926, and the Regent Theatre in Brisbane in 1929 (Ahern 1979, pp. 17–19). The Carrier Company established an office in Sydney in January 1932, which designed a number of installations, mostly industrial. In 1938 a factory was built for the local manufacture of equipment. Installations before World War II included a few office buildings and factories, the operating theatre at the Prince Henry Hospital in Sydney, the entire Broken Hill Hospital, the original Sydney Water Board Building, and parts of the David Jones Department Store and of the Hotel Australia in Sydney (Carrier 1982, p. 6).

When World War II ended, air conditioning was thus still restricted to a few buildings, even in America. One reason was the limited range of refrigerating fluids available before the war; all were poisonous, and a vapour leakage was not preventable with complete certainty. The freons, a group of stable and physiologically harmless fluoro-carbon fluids, had been developed in the 1930s. Freons have since been found to have a potential for depleting the ozone layer, and they are being replaced by the HFCs (hydro-fluoro-carbons).

The wartime advances in air-conditioning technology were largely due to research associated with its use in US Navy ships operating in tropical waters. Some of the modifications to the air conditioning of American submarines were made in Australia (Carrier 1982, p. 8).

After the war the use of air conditioning increased greatly in America. This was partly due to the technical advances, and partly to Modern Architecture, which had been an 'alternative' approach between the wars, becoming the new orthodoxy in America and in Europe. Curtain walls containing a great deal of glass were used increasingly in place of masonry walls, and with these summer air conditioning became a necessity instead of a luxury.

The new American architecture was studied with great interest in Australia. At the 1954 annual conference of the Royal Australian Institute of Architects the auditorium was dominated by a huge picture of Wallace Harrison's glass-walled United Nations Building, and there was much discussion on the use of dry lightweight construction, and on the desirability of removing the 42-year-old restriction on the height of buildings (see p. 76). In 1955 the 150-foot limit was lifted (see p. 188).

The effect on building heights and on the use of air conditioning was not immediate, but when it came it was far reaching. In 1957 there were only three fully air-conditioned office buildings in Sydney. Ten years later it had become difficult to let an office suite in the Central Business Districts of Sydney and Melbourne that was not air conditioned.

## A constant thermal comfort criterion?

Thermal comfort in cold weather is determined by the (dry bulb) temperature if one is sufficiently distant from an open fire or a radiator. The air may get too dry as it is heated, but this can be corrected by allowing water to evaporate from a dish.

In warm weather humidity and air movement play an important part. In the early 1920s the American Society of Heating and Ventilating Engineers, one of the societies that later merged into ASHRAE, initiated a major investigation at its research laboratories in Pittsburgh (Houghten & Yaglou 1923, p. 163). The people taking part in the experiment were asked what combinations of temperature, humidity and air movement gave the same sensation of comfort when they were engaged on light physical activity, and which were the conditions that gave the greatest comfort.

In the late 1950s ASHRAE decided to repeat the experiments at its laboratories, now in Cleveland, to allow for changes in clothing and in environmental perception in the interim. In 1960 the ASHRAE laboratories were closed, and the work was transferred to Kansas State University as a joint KSU–ASHRAE investigation, which involved 1600 students (ASHRAE 1977, pp. 19–28). P. O. Fanger, who had been a member of this research team, continued the study, after his return to his native Denmark. He included old and young people, who had been missing from the American tests (Fanger 1970, p. 85). These investigations are indirectly the basis also of Australian air-conditioning design.

There is no reason to query the accuracy and thoroughness of this research, but it is questionable whether a constant temperature criterion is best for summer comfort. It may be for a person leading an interesting life, who remains all day within this 'well-tempered environment' (Banham 1969, p. 11). He leaves his air-conditioned home via his air-conditioned garage to drive his air-conditioned car to the underground garage in his air-conditioned office. The process is then repeated for lunch, for recreation in the evening, and eventually for going home.

This lifestyle is possible for very important people (Kempton & Lutzenhiser 1992, p. 2). Air conditioning is less agreeable for those who have to go outdoors to go to work, to get lunch, and to transact business in other buildings (de Dear & Auliciems 1988, p. 2). This causes little difficulty in winter, when one can put on an overcoat on a cold day, but it creates comfort problems on a hot summer day.

Instead of maintaining a constant 'optimum' temperature, it would be more sensible to adjust the indoor temperature according to the conditions outdoors. On a hot day the indoor temperature could be allowed to rise, although not as much, to reduce the contrast between indoors and outdoors; this would incidentally save energy.

The temperature changes not merely with the season, it also changes during the day. Most air-conditioned buildings ignore this diurnal change. It is, however, part of our biological rhythm, and part of the pleasure we get from being outdoors on a nice day. There is a case for diurnal temperature movement in air-conditioned buildings, allowing the temperature to rise slightly during the day, and dropping it again in the afternoon, which would save energy for the cost of a simple control. There may also be a case for some parts of the building being warmer or cooler than the rest to provide a contrast for the occupants, the equivalent of a sunroom or a shady porch.

The variation in temperature during the day would break the monotony for the large number of people in any building who have to stay inside all day doing a boring job. One often hears complaints about 'sick' buildings (de Dear & Auliciems 1988, p. 23). Undoubtedly there are buildings where malaise is caused by failure to clean air filters; however, many complaints are due to boredom with the job, and a change in the uniformity of the environment would help. This might give air-conditioned buildings some of the environmental characteristics of the best buildings surviving from the late nineteenth and early twentieth centuries, and it would also save some energy.

## Energy conservation and solar energy

The oil-producing countries trebled the oil price and restricted its sale in 1973–4 (see p. 123). By this they created an energy crisis, but in retrospect they probably did the oil-importing countries a good turn, although that was not the intention. The Americans in particular reacted promptly (ASHRAE 1975; HUD 1975), by initiating research programmes to determine how energy consumption might be further reduced, and what alternative energy sources might be used in place of oil. Several European countries followed suit. Australia was a little slower to react (NEAC 1977); because of the milder climate in its big cities it used less energy for buildings, and it was not a major importer of Middle Eastern oil.

These investigations revealed that energy had been used wastefully because it was so cheap and plentiful in the 1960s, and there was plenty of scope for conservation (for example, Cowan 1983, pp. 9–12) by better energy management and programming, by better insulation, and in some cases by improved air-conditioning equipment.

The search for alternative energy sources was less successful. The use of solar energy had initial support not merely from the government and building industry professionals, but from the alternative lifestyle groups that gained prominence in the late 1960s and 1970s, who opposed the use of atomic energy. Solar energy is safe, clean and free, but its intensity on the Earth's surface is very low, and concentrating it can be expensive.

Its most efficient use is in hydroelectricity, which depends on the sun evaporating water at a low level so that it can be deposited by rain at a higher level; but as this had been used for more than half a century, it was not usually included in the discussion on solar energy. Other indirect methods, such as wind energy, tidal energy, and conversion of fast-growing plants into industrial alcohol, worked well from an engineering point of view, as did direct methods, like concentrating solar radiation with parabolic mirrors, tracking the movement of the sun; but the electricity produced cost more than conventional electricity in Australia.

Solar water heating, which had been used since the 1940s (Daniels 1964, pp. 104–22) was more successful, and it had been largely pioneered by Australian manufacturers. It needs to be supplemented by conventional heating, unless one is willing to take a cold shower on overcast days. A substantial part of Australian water heating is done by off-peak electricity, which uses the output of the power stations during the hours when the general demand is low. Before 1973 most electricity companies refused to allow off-peak rates to people who installed solar water heaters, on the economically justifiable ground that they would use the electricity only at irregular intervals. This rendered solar hot water more expensive than that from electric off-peak heaters. Some companies still retain this restriction. As solar water heating is one of the few presently economic uses of solar energy, it would be appropriate to remove the restriction where it remains.

One aspect of solar energy that has worked well is passive solar energy. The term was coined by an American applicant for a research grant, who wanted to explore the use of sunlight penetration in the classical and Renaissance era, which had not previously been thought of as a use of solar energy. He thought that he would stand a better chance if he used the term 'solar energy'.

Passive solar energy has since become a legitimate branch of solar energy research. It is often economical because it does not require any expensive infrastructure or equipment, merely a modification of the building. Its most successful form is the 'direct gain' method (Figure 64), which is essentially the method advocated by Socrates (see p. 124). It was used in various countries throughout the centuries, including Australia from about the middle of the nineteenth century. With scientifically calculated roof overhangs or sunshades, and with interior surfaces designed to store the solar radiation that penetrated, in many parts of Australia it was sufficient for winter heating on sunny days until several hours after sunset.

There are a number of passive solar methods (Cowan & Smith 1983, pp. 88–95) which are thermally more efficient than the direct-gain method, but these involve additional cost. The building of a solar greenhouse is justified if one has use for an overheated room in winter, and that of a roof pool (whose surface would have to be covered or exposed according to a schedule) if one likes a swimming pool on the roof rather than one in the ground. There are also numerous proposals for the use of heat/coolness stores, which involve the opening or closing of ducts or shutters at definite times twice daily. An owner willing and able to attend to this small task when it is required can gain additional thermal comfort, and/or conserve energy; a servo-mechanism installed for the purpose is unlikely to be cost effective.

While solar energy has so far failed to make the contribution to energy conservation that had been expected in the late 1970s, the extensive discussion of it has drawn attention to the important effect that the sun has on the interior thermal environment of buildings, and how it can be used to advantage by relatively simple constructional measures.

However, this may change by using photovoltaic panels to convert solar radiation directly into electricity. They have been employed for many years to generate power for spaceships. Initially the cost of the electricity generated by photovoltaic panels was more than a thousand

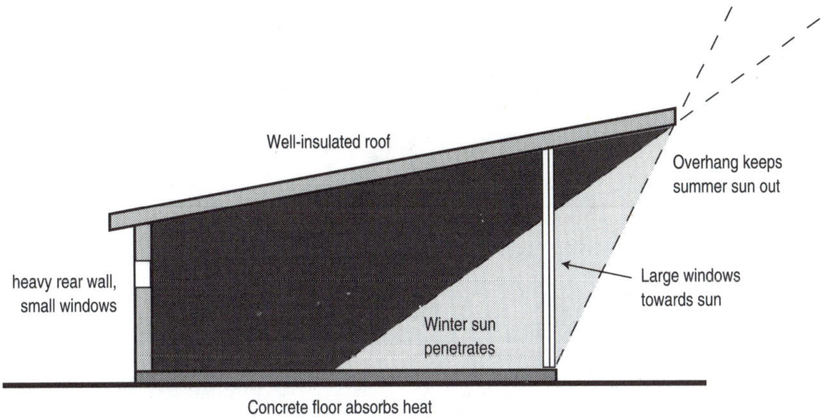


**Figure 64** The direct-gain method of passive solar energy. The roof overhang or sunshade permits penetration of solar radiation in winter, but excludes it in summer. The winter heating is more effective if solar radiation can be stored in a floor with a high thermal inertia; this would preclude the use of an insulating layer, such as a thick carpet.

(Drawing by Peter R. Smith)

times as high as that of conventional electricity; but the margin has now been reduced to a few per cent.

Sufficient photovoltaic panels will be installed in the buildings of the Olympic Village for the Sydney 2000 Games to supply their need for electric power. The panels will feed electricity into the grid on sunny days, and draw from it when the sun is overcast. The objective is to emphasise the 'environmentally friendly' design of the buildings, and it is expected that the additional cost will be modest.

At the time of writing it seems possible that photovoltaic electricity in the foreseeable future will be able to compete with electricity produced from fossil fuels, even in metropolitan areas.

# 7

# WATER AND FIRE SERVICES

## Water supply and sewage disposal

Several ancient civilisations had a piped water supply and a waste disposal system on a small scale for a temple or the palace of the ruling family. The Roman Empire, however, established a near-universal water supply and an extensive sewage disposal system in its capital and in many of its cities. The City of Rome eventually had eleven aqueducts, the first built in 312 BC and the last in AD 226. These had a combined length of 560 km. They were mostly underground conduits; only about one-tenth was above ground, carried on arches that are still conspicuous in the landscape. The Roman aqueducts, unlike modern water mains, ran at atmospheric pressure, and they needed a steady fall for successful operation. After the collapse of the West Roman Empire in the fifth century the aqueducts and sewers in its region gradually fell into disrepair.

Even though several aqueducts might be built on top of one another as they came close to the city, the water in the various aqueducts was kept separate, because it was not all of the same quality; only some was pure enough for drinking water. Frontinus, who was for many years the prefect of the water supply, wrote a book in AD 97 *De Aquis Urbis Romae* (About the water supply of the City of Rome) (Winslow 1965, p. 17) in which he explained that the best way to obtain a supply of pure water was to get it from a pure source; the Romans were unable to purify water. This is still good advice today to minimise the need for purification, and it is followed by most Australian water authorities. Fortunately there are enough sparsely inhabited hillsides, that can be kept largely free of habitation and grazing animals, to act as catchment areas. Many European cities presently lack this resource.

Frontinus estimated the daily water consumption of Rome in his day, when there were only nine aqueducts, as 24 360 quinariae, a measure based on the cross-section of the ducts and ignoring the velocity of flow. Professor Isaac (1958, p. 16) estimated the flow velocity from the average inclination of the aqueducts, and concluded that the water supply of Rome was about 11 cubic metres per second. This was about 7½ times that required by his native Newcastle-upon-Tyne and Gateshead, which at that time had a population of about 350 000, which he estimated to be about the same as the population of Rome in Frontinus's day. Others have arrived at somewhat different estimates of the quantity of water and of the population of Rome, but there is no doubt that the water consumption per person in Imperial Rome was much higher than it is today.

The main reason was that the Romans did not use any watertaps, although they probably had the technical capacity to make them. The water flowed continuously from the fountains and standpipes, from which water was collected by people, except for a few who had a piped supply. The water was not entirely wasted; it was recycled for other uses, the last being for flushing the public toilets. These consisted of slabs, generally of marble with large oval holes; underneath there was a continuous stream of running water. Several of these toilets were buried in Pompeii and in Ostia, and were found intact during excavations.

In 1842 Edwin Chadwick was employed by the British Royal Commission on the Poor Laws as an investigator, and his *Report on the Sanitary Conditions of the Labouring Classes of Great Britain* (Chadwick 1842) resulted in far-reaching reforms. Chadwick visited the Roman ruins in Italy as part of his investigation, and he concluded that the general sanitary standards of ancient Rome had been greatly superior to those of Britain in his time.

Many centuries after the fall of Rome a piped water supply was gradually restored. By the eighteenth century in cities such as Paris (Figure 65) and London it was brought into the houses of the wealthy, and to standpipes or fountains for the general populace. Water carriers were available to deliver it to houses that had no piped supply. The fact that the water was piped did not mean, however, that it was pure. Indeed, in the big European cities this was unlikely at the time the First Fleet set sail.

Even today people in some countries drink dirty and bacteria-infected water. A proportion of the children die as a result, and the survivors develop a measure of immunity. A visitor who has not acquired this immunity is liable to be taken ill or die after drinking the infected water. This was the position in a large part of the world until the nineteenth century, before an increased understanding of infectious diseases led to drastic reforms.

In cities situated on rivers, such as London, the problem was largely caused by the long-standing custom to discharge sewage from nearby houses into the river, and also to pump drinking water from the river. Where spring water was used, the spring might be polluted by a nearby latrine-cesspit.

In the country it was easier to dispose of refuse without pollution; but the water might still be bacteria-infested due to ignorance of the

**Figure 65** Water pipes in Paris.
(From C. M. de la Gardette, *L'art du plombier et du fontainier*, Paris, 1773)

causes. George Goodwin, who edited *The Builder* for many years, in 1859 published *Town Swamps and Social Bridges*, in which he illustrated a village, delightful from a distance, but in close-up he showed that the water pump was next to a cesspit and to a pigsty, and the caption explained that there was a cemetery between the pump and the spring that fed it.

There are two aspects to the problem of sewage disposal. The first is to get the faeces out of the building. The original method, which persisted in parts of Australia into the second half of the twentieth century, was to use a bucket, which was removed by a carter once or twice a week.

In the late seventeenth century the water closet (WC) was invented, apparently in England, because in France it was referred to as *cabinet d'aisance à l'anglaise*. The original device, known as a pan toilet (Figure 66a) was used throughout most of the eighteenth century, but it had a number of moving parts that were difficult to clean, so that the problem of the bad smell remained. In the late eighteenth century Joseph Bramah, an inventor of many ingenious devices (see also p. 173), patented the valve toilet (Figure 66b), which still had a movable part, but this was easily accessible. This type of WC is still used in some Australian railway carriages.

The washout toilet (Figure 66c) was developed in the 1870s by an English pottery manufacturer, Thomas William Twyford. As it consisted entirely of glazed pottery and had no moving parts, it was easy to keep clean, and it has been universally accepted.

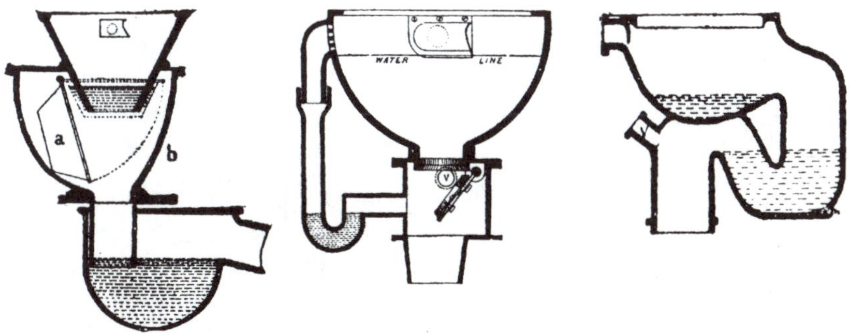


(a) (b) (c)

**Figure 66** Water closets:
(a) pan toilet;
(b) Bramah valve toilet;
(c) Twyford's washout toilet.

(Illustrations from the article, 'Sewerage' by Sir Alfred Ewing in the *Encyclopaedia Britannica* 9th edn, Adam & Charles Black, Edinburgh, 1886, vol. 21, pp. 716–7)

The first Australian WC was probably installed in the First Government House (see p. 19) during Governor Hunter's term of office (1796–1800).

However, the installation of a water closet did not solve the problem of sewage disposal, it merely avoided the need for the unpleasant job of handling the 'nightsoil'. In the eighteenth century the faeces were removed by the flush to a soaking pit or to a nearby stretch of water.

The disposal of the sewage of an entire municipality required a large and expensive public works project. Hamburg, then still a Free City State, was the first to build one in 1842. The London Main Drainage Scheme was started in 1856, and it became the prototype for the difficult problem of disposing of Melbourne's sewage (see p. 160).

Once the sewage was removed, it became possible to produce a clean water supply from a river or from groundwater. However, the purity of the drinking water remained a problem until the early twentieth century, and some houses installed additional filters (Figure 67).

## Water supply in Australia

The prime need for any settlement is a permanent supply of water. Governor Arthur Phillip examined the different coves of Sydney Harbour and selected the one that had the best spring of water.

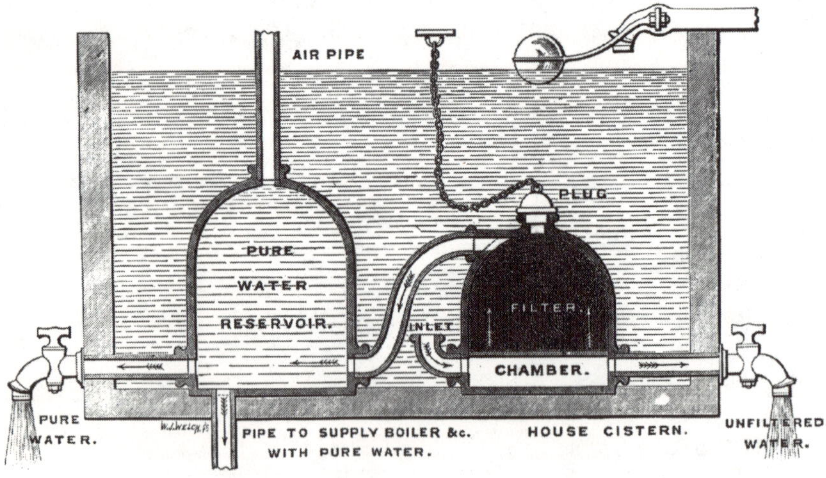


**Figure 67** Water filter for installation in a water-storage cistern.

(Illustration from the article 'Filtration' by A. B. McDowall in the *Encyclopaedia Britannica* 9th edn, A. & C. Black, Edinburgh, 1879. vol. 9, p. 169)

A further small stream was found running into the harbour at Rushcutters Bay; it is still there. However, the Tank Stream (see p. 18), now an underground stormwater drain, remained the main source of the Sydney water supply for more than thirty years (Aird 1961, p. 1). Springs rising from the sandstone came to the surface in marshy ground west of the present Hyde Park between Market and Park Streets. They turned into a clear stream between Market and King Streets, and reached the high-tide level of Sydney Harbour just below Bridge Street.

As the Tank Stream became more and more polluted by sewage and other wastes, a number of springs were fitted with pumps to provide cleaner drinking water (see p. 18).

By 1820 water was being brought by watercart from the Lachlan Swamps, the present Centennial Park. In 1824 John Busby, a mineral surveyor newly arrived from England, surveyed a route for a tunnel to deliver water from the Lachlan Swamps to a 15 million gallon (70 million litres) reservoir at the Racecourse (now Hyde Park). The population of Sydney was then about 10 000.

Work started on 'Busby's Bore' in 1827, using convict labour. It was not finished until 1837, but by 1833 enough water was running in it from seepage along the completed part of the tunnel to deliver some water to the reservoir.

The tunnel was cut through the solid sandstone, and in a few places, where the excavation was through sand, it was lined with sandstone masonry. It was similar to a Roman aqueduct, but the standard of workmanship was well below that of the surviving Roman work. The tunnel rises and dips, and it zigzags horizontally. Its cross-section varies from 3 feet × 3 feet (1 m × 1 m) to 10 feet × 11 feet (3 m × 3 m). This was blamed later on the fact that Busby never went down into the tunnel to inspect the work, because he was afraid of violence from the convict workmen (Aird 1961, p. 7). However, Busby, unlike most other English migrants, was already fifty-nine years old when he arrived and seventy-two when he finished the bore. Few men of his age would today be able to go down the small manholes, some of which still exist, that led into the bore, or crawl through a 1 m × 1 m passage.

A pipe was built out from the Hyde Park Reservoir on pylons; it was at its outlet high enough for a watercart to drive underneath to be filled. The outlet is now marked by a small commemorative fountain. This was initially the main method of distribution. However, for a fee of ten shillings householders could be connected to the water supply, and in 1844 reticulation pipes were laid from the bore to various parts of Sydney. In addition hand-pumps were built along the bore so that people could draw water. One of these pumps survives outside the Victoria Barracks in Oxford Street (illustrated by Kennedy & Kennedy 1993, p. 30).

The bore was the main source of water until the Botany Swamps Scheme came into operation in 1859. The water from one, and later several, storage dams was pumped through a 30 inch (760 mm) main by a 100 HP (75 kW) steam-powered pump to a new reservoir in Crown Street, which is still in use. As there was no telegraphic communication, a standpipe was erected outside the Victoria Barracks at the same level as the top of the reservoir; this could be seen from Botany with a telescope, and it indicated when the reservoir was filled.

By 1867 it had become clear that allowance had to be made for a number of consecutive dry seasons, because they were a regular part of the local climate. The population of the Sydney region now exceeded 100 000, and Governor John Young set up a commission, which included Edward Moriarty, the Engineer-in-Chief of Harbours and Rivers in the Department of Public Works, and John Smith, the Professor of Physics at the new University of Sydney. In 1869 this Commission recommended the Upper Nepean Scheme, which led eventually to the construction of the Prospect Reservoir, and of five large storage dams on the Nepean River and its tributaries. The Prospect Dam was completed in 1888, the Cataract Dam in 1907, the Cordeaux Dam in 1926, the Avon Dam in 1928, the Nepean Dam in 1935, and the Warragamba Dam in 1960. In 1941 a dam was completed across the Woronora River, which is not part of the Hawkesbury–Nepean river system, to serve the southern suburbs of Sydney.

Aird (1961, p. 106) points out that the water storage provided for the population of Sydney (220 000 gallons (1 000 000 litres) for each person served) is much higher than for London (4000 gallons per person) or New York (55 000 gallons per person). This is partly due to the relatively high consumption in Australia, but more particularly to the long periods of drought; Sydney storage was determined from observations of the rainfall during the 1932–42 drought.

With the completion of the new dams stored water was available for distribution over a much wider area, and the distribution system increased greatly in complexity towards the end of the twentieth century. Although some water channels between reservoirs were still running at atmospheric pressure, as they had been since the days of Rome and in Busby's Bore, the distribution mains were running full under pressure supplied by the head of the reservoir. For the higher parts of the metropolitan area 'rising mains' were used, which ran uphill, with the necessary pressure supplied by pumps.

Parramatta had a water supply system from 1815, when a well was sunk for public use. In 1818 Governor Macquarie ordered the building of the Town Dam (initially of timber) at the foot of Marsden Street, approximately where the fresh water in the Parramatta River met the salt water. Another dam was built 1½ km upstream in 1828. Distribution was initially by watercart. In 1856 a dam was built further north on Lake Parramatta.

The only other part of the Sydney Metropolitan Area that developed an independent system was Manly, but that was not until 1892.

In 1888 the Board of Water Supply and Sewerage (later called the Metropolitan Water Sewerage and Drainage Board, now Sydney Water) was formed to operate the existing Sydney water supply and sewerage systems, and it also took over the Manly system in 1902 and the Parramatta system in 1914.

Some houses too remote from the Board's mains remained outside the system, and these obtained their water either from a well, or by collecting the rainwater falling on their roofs in large tanks.

Melbourne relied for its water supply on wells and on the Yarra River until the newly appointed Board of Commissioners of Sewers and Water Supply built the Yan Yean Reservoir in 1853. This was formed by a dam across the Plenty River, a tributary of the Yarra, about 30 km north of Melbourne. It provided an adequate supply for many years, and remained the main source of drinking water until the 1920s when the Maroondah, O'Shanassy and Silvan Dams were started, the last finished by 1932. After World War II the Upper Yarra Dam was built in the 1950s, the Cardinia Dam in the 1960s and the Sugarloaf Dam in the 1970s.

Water supply for human consumption (as distinct from agriculture) has not been a major resource problem in the coastal regions of Australia, where most people live; although financing the necessary public works has at times required considerable political skill. Compared to England, there is still plenty of open country; most towns have potential catchment areas nearby where water could be collected that requires only a minimum of treatment, and although allowance has to be made for long seasons of drought, the average rainfall in most coastal regions is comparatively high.

The problem is very different in the arid areas. Alice Springs owes its location to the springs; in some other centres of habitation, water has been obtained from wells, even in regions with very low rainfall.

However, the location of arid-zone settlements is often determined by mineral finds, and it may be necessary to provide a water supply by pipeline. The most notable example is the system built for the goldfield towns of Coolgardie and Kalgoorlie, initiated by Charles Y. O'Connor, the Western Australian Government's Chief Engineer. Work started in 1898 on a reservoir in the Darling Range near Perth. The water was then transported 500 km by pipeline and lifted 400 m. This was accomplished by eight pumping stations between the reservoir and Coolgardie; from there the water flowed by gravity to Kalgoorlie.

There were many delays due to the unusual nature of the work, and the need to import most of the engineering equipment from England (Carroll 1988, p. 104). A great deal of the public criticism of the project as impracticable and a waste of public money was directed against O'Connor, who had had a distinguished career as an engineer in New Zealand and in Australia. In March 1902 he committed suicide. The pipeline began to flow in January 1903, and it is still in use.

## Sewage disposal in Sydney and Melbourne

There are four principal ways of disposing of faeces:

1 To build a toilet over a cesspit, a hole in the ground: the sewage may drain away slowly, and this may affect the water supply if it relies on ground or spring water. This has been a major cause of the spread of infectious diseases.

   If the soil is relatively impervious, the hole will gradually fill up. It is then filled in, and the toilet is moved. This method was used extensively in the early days, and it is still used in the country. In a built-up area it is today considered unacceptable.

2 To build a septic tank in which bacteria feed on the sewage: the comparatively innocuous effluent can then be discharged into the garden where it acts as a fertiliser, or it can be tanked away for disposal. Septic tanks were widely used from the late nineteenth century (Figure 68), and they are still used today where a connection to the sewerage system is difficult, and where there is a garden large enough to absorb the effluent. The design is controlled by regulations that differ in each state (Sharp 1969, pp. 324–34).

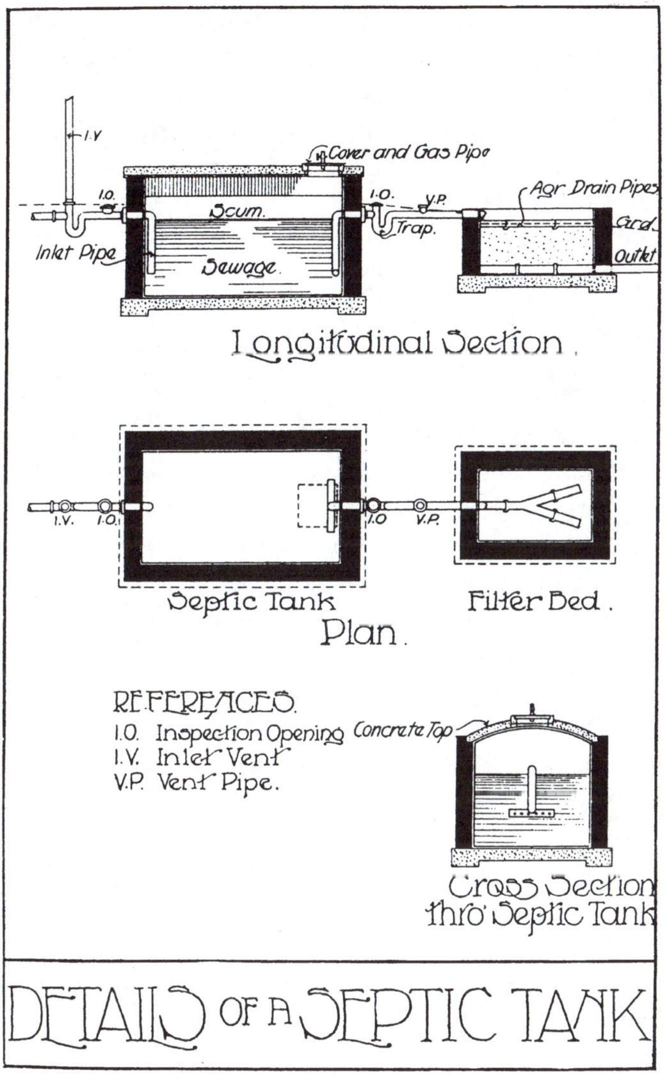


**Figure 68** Septic tank for use in Australia.
(Illustrated by Haddon [1908, p. 449])

3 To deposit the faeces in a bucket that is removed once or twice a week by a carter: this is a most unsatisfactory system, both for the building owner and the carter, but it was widely used. In Sydney almost one-third of the suburban houses were still dependent on sewage carters in 1960, when a determined effort was made to extend the piped sewerage system to the entire metropolitan area.

4 To flush the faeces into a pipe that takes it to a disposal plant, or deposits it in the ocean: this requires a network of pipes, and generally also pumping stations, and it may require a purification plant. This is the solution to the problem that is favoured today. In general, it requires more money than a water-supply system. A city cannot survive without a water supply, but it can exist without piped sewage disposal, although its quality of life is greatly reduced.

A few houses in Sydney, including the first Government House (see p. 19) probably had water closets in the early nineteenth century, from which the faeces were water-borne to a soaking pit well clear of the house, or to Sydney Harbour if it was close enough.

In 1855 the Sydney City Council commenced construction of the first Australian public sewerage system. It was completed in 1857, which was early by world standards. The liquid from the sewers under the main north–south streets was collected by a cross-sewer (Figure 69), which discharged into a 'treatment plant' at Bennelong Point (where the Opera House now stands). This allowed the solids to settle, and they were eventually sold as manure. The remaining liquid was discharged untreated into Sydney Harbour.

The oval-shaped brick sewers are remarkable for an early use of imported English 'Roman' cement (see p. 8), which was used to make the brickwork waterproof, instead of the lime mortar that had been previously employed. The sewer-vent in the form of an obelisk at the junction of Elizabeth and Bathurst Streets (illustrated by Thorpe 1956, p. 57) was built as part of the scheme.

In the following year a subsidiary system was built in the Rocks, which discharged into the recently completed Semicircular Quay at Queens Wharf. The Tank Stream had become the city's principal stormwater drain (Figure 69), and it also discharged into Semicircular Quay.

In 1873, after a severe outbreak of typhoid fever, a specially appointed Sewerage and Health Board recommended that the practice of flushing WCs directly from the waterpipes be discontinued, and that each WC have a separate flushing cistern with an air gap to avoid any possibility of contamination by backfeed. It also proposed that a new outfall sewer be built to intercept most of the existing sewers draining into the Harbour and take the discharge to an ocean outfall. Two separate schemes were drawn up for a northern system with an ocean outfall

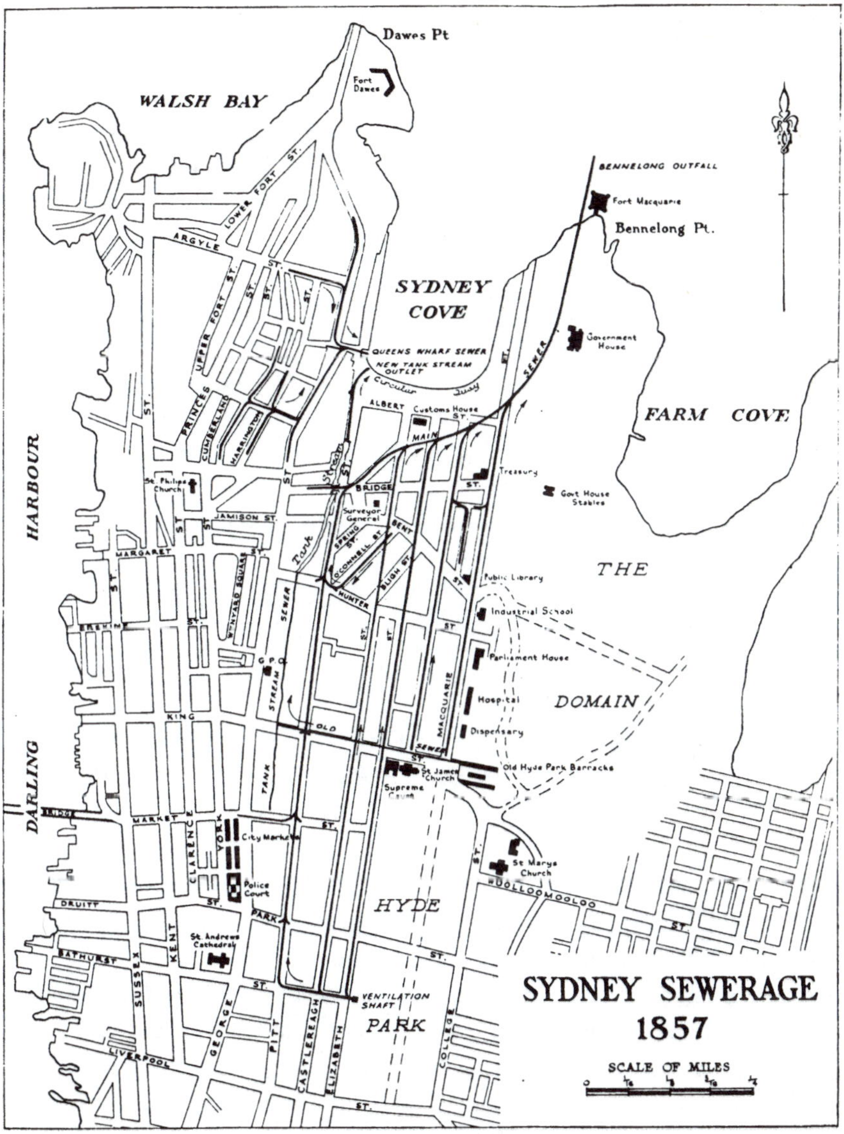


**Figure 69** The public sewerage system built by the Sydney City Council 1855–6. The sewers running under Pitt, Castlereagh, Elizabeth, Phillip and Macquarie Streets had an outfall beyond Bennelong Point. The small system in the Rocks area had its outfall at the Queens Wharf in Semicircular Quay. The Tank Stream had become a stormwater sewer, but it was still an open stream; the part from Hunter Street to Bridge Street was put underground in 1860.

(Reproduced from Aird [1961, p. 134], with Sydney Water's permission. Copyright vested in Sydney Water Corporation Ltd)

at Ben Buckler, Bondi, and a southern system terminating at a sewage farm near Botany Bay.

These proposals were implemented by the Board of Water Supply and Sewerage formed in 1888. The northern system included the two Annandale sewerage aqueducts (see p. 90). In 1904 the sewage from low-lying areas, which still went into the harbour, was lifted by electric pumps through rising mains, using power from the Tramways' Ultimo Power House (Figure 50).

The sewage farm for the southern system led to complaints after several years of operation, and in 1916 the sewage was discharged through an ocean outfall from Botany (now Sydney Airport) to Malabar (Long Bay). A third system for the northern suburbs with an ocean outfall at North Head was built between 1916 and 1927.

As the population of Sydney grew there were increasing complaints about sewage from the outfalls, which were only a few metres beyond the low-water mark, being washed ashore. Sewage treatment plants were installed, which gave only primary treatment, and when these failed to solve the problem three new ocean outfall tunnels were built in the 1980s that discharge the effluent through a series of diffusers approximately 4 km offshore.

Melbourne did not acquire a sewerage system until 1898. The gold rush of 1851 had caused a rapid increase in the population of Victoria. In 1851 it was 77 000, in 1854 237 000, and in 1857 410 000 (Cannon 1993, p. 1). About a quarter settled in Melbourne. By 1880 Melbourne had a population of almost half a million (Dingle & Rasmussen 1991, p. 27). It was the third-largest city in the British Empire (excepting India) in population, and the second-largest in surface area. It had become the largest and richest city in Australia. It had the finest buildings, and the headquarters of the most important banks and commercial companies. In 1853 it built a good water-supply system, and by 1880 almost every cottage in Melbourne had a bathroom, at a time when bathrooms were still a rarity in British working-class cottages. Melbourne's water consumption per person was roughly double that of London. However, there was still no sewage-disposal system, and the city depended mainly on sewage carters.

As a result waterborne diseases were prevalent. In 1889, after a virulent typhoid epidemic, the Victorian government appointed a Royal Commission, which led to the creation of the Melbourne and Metropolitan Board of Works.

After considering several alternative proposals by a visiting British consulting engineer, James Mansergh, the Board's engineer-in-chief, William Thwaites, decided on a single system discharging into a sewage farm at Werribee. It had two advantages over Sydney's unsuccessful sewage farm at Botany Bay: it was well beyond the metropolitan area, and it had a much larger land area at its disposal, 8847 acres (36 sq. km).

Werribee operates on two principles, land filtration and grass filtration. It has a low rainfall evenly spread over the year, the land is level, its soil is suitable for filtration, and the effluent can be discharged into adjacent water. It is one of the world's largest remaining sewage purification areas. Most other cities that used them gave them up because they occupied too much land.

The sewerage system designed in 1889 required the construction of 58 km of mains and of 440 km of reticulation sewers, and a pumping station at Spotswood, where four 200 kW steam-driven pumps lifted the sewage to a level where it would flow through the 21 km main outfall sewer to Werribee by gravity. The number of pumps was later increased, and starting in 1921 they were converted to electricity (Dingle & Rasmussen 1991).

The Werribee plant was opened in 1898, and within twenty-five years the number of houses not connected to sewers in Melbourne had become less than that in Sydney. The sewage farm proved adequate until the 1960s, when a new South Eastern Sewerage System was built at Carrum, on the opposite side of the metropolitan area.

## Fire fighting

When the First Fleet sailed for Botany Bay, sanitary conditions in England were still primitive. Fire fighting, however, had already reached some degree of sophistication.

The Great Fire of London in 1666 had resulted in major reforms to building regulations to limit fires, and in the setting up of fire brigades by insurance companies. These were small and inadequately equipped, but they were the first fire brigades since the Fall of Rome. Insurance companies later also initiated the first effective fire-fighting services in Australia.

Whether a fire brigade during a big fire could and should destroy buildings to make a fire break has always been a controversial issue. In ancient Rome the Prefect of the *Vigiles* (fire fighters) was equal in rank to the Prefect of the Praetorian Guard, and he had the necessary authority, which he used if a fire got too big; it was his most effective method of stopping a great fire.

By the nineteenth century most British cities had councils, generally elected by property owners. In London this had already been so since medieval times, and the Lord Mayor in 1666 was reluctant to order the demolition of buildings to make a firebreak before the fire got out of hand.

This was also a controversial issue in Australia. When the first Bill to establish a New South Wales Fire Brigade was introduced in 1854, it included provision for a 'fund for the compensation of all persons whose properties may be destroyed . . . by authorized personnel . . . to arrest the progress of fire and in cases of great danger'. The question of non-fire property damage caused much debate, and the Bill did not pass a Second Reading.

In the eighteenth and early nineteenth centuries the principal method of stopping a fire was to douse it with water. The *City of Sydney Building Act* of 1837 permitted the Police Magistrate to erect 'stopcocks and firecocks' on water mains in any street or pavement; but there would have been few such mains until ten years later. The principal method of water supply for fires was therefore from watercarts or buckets, until the piped water supply improved.

Water supply for fire fighting remained a major problem until 1889, when the newly created Metropolitan Board of Water Supply and Sewerage (see p. 155) replaced the old-fashioned fire plugs in the City with about 1500 screw-down hydrants, and in the suburbs with ball hydrants. This was followed by the gradual replacement of the 3 inch and 4 inch mains by 6 inch (150 mm) mains in high-density and industrial areas, installed at greater depth to avoid fracture in case of falling walls (Adrian 1984, p. 59).

Handpumps had been used for fire fighting since medieval times. In the mid eighteenth century a two-cylinder handpump was developed in England, which could project a *continuous* jet of water, although with limited pressure. In the early nineteenth century the first steampowered fire engine, still horse-drawn, was built in America (Figure 70). The 'steamers' could project a high-powered jet of water, and greatly improved the effectiveness of the fire fighters; the first was imported into Australia (from England) in 1865.

**Figure 70** 'Steamer' fire pump made in America in the 1870s. The engine cylinder had a diameter of 7½ inches (190 mm) and a stroke of 8 inches (200 mm). The tubular boiler provided a capacity for 500 strokes per minute, working at a pressure of about 90 psi (620 kPa). The engine weighed 8000 lb (3600 kg), and was drawn by two horses.

(Illustration for the article 'Fire' by General A. P. Rockwell in the *Encyclopaedia Britannica*, 9th edn, A. & C. Black, Edinburgh, 1879, vol. 9, p. 236)

The first fire engines were acquired by Governor Bourke in 1822 for use by the Army. However, there are no detailed descriptions of these or subsequent fire engines until 1851. A photograph of the principal fire engine of the Insurance Companies Fire Brigade, taken in 1851, has been published by Adrian (1984, p. 17); it shows a watercart with a single-piston pump, drawn by one horse.

The first three steamers, imported by Insurance Companies in 1865, 1871 and 1874, were transferred to the newly-formed NSW Fire Brigade in 1884. In 1892 it imported a more powerful steamer from England, which operated at 100–125 psi (690–860 kPa) and was capable of delivering 1000 gallons (4600 litres) of water per minute to a height of 200 feet (60 m). After the purchase of a second steamer in 1899, it was decided to give a demonstration of 'cleaning' the tower of the Sydney Town Hall. Adrian (1984, p. 160) reproduces an illustration that shows the water jets reaching well above the top of the tower.

These steamers were still horsedrawn. A machine was developed that dropped the harnesses on the horses for a speedy turnout. A piped gas supply had been available in Sydney since 1846 (see p. 113). When a fire engine returned to the station, a kerosene-soaked tinder of wood and coal, fitted in a box, was placed in the fire box, and the engine was positioned slightly behind a gas outlet that projected from the floor. A pilot light constantly burned at the head of this outlet, and when a lever was pulled, a sudden blast of ignited gas flashed upwards to light the tinder, and started the steam engine (Adrian 1984, p. 166).

However, hand-operated pumps pulled to the fire by teams of four to six firemen continued in use until well into the twentieth century at the smaller fire stations. They were reliable, and easy to use by part-time firemen.

In the early twentieth century the horses were gradually replaced, partly because the Boer War had created a shortage of suitable horses. Melbourne sold its last horse in 1918; Sydney did not do so until 1927. The original intention had been to use steam for propulsion, and consideration was also given to the use of the recently developed electric accumulators; but when the Sydney brigade acquired the first self-propelled fire engine in 1906, it was petrol-driven; it also provided the power for the water pumps.

The length of the earliest ladders used by firemen was limited by their transportability. They were replaced by extension ladders, which could reach up to 80 feet (24 m). In 1893 Sydney imported from England the first manually operated and horse-drawn turntable; it also had a reach of 80 feet (24 m). Subsequent turntables were locally made, some in the Sydney Fire Brigade's workshops. The first electrically operated ladder was produced in 1918 in the Fire Brigade Workshop from one that had been manually operated.

Firefighting in Sydney was originally the responsibility of the people at large, and specifically of the Army; the first primitive fire engines were housed in 1822 in the Wynyard Military Barracks. In 1837 the Australian Insurance Company, established the previous year, started to write fire-insurance policies, and it supplied its clients with cast-iron plaques to display on their houses. It employed eight wharf labourers as part-time firemen, and established two depots with buckets, ladders and axes for their use.

In 1837 the Alliance Insurance Company also established a small 'fire brigade', and so did the Mutual Fire Insurance Association in 1841. Eventually there were almost a dozen competing fire-insurance companies in Sydney. Each produced a different brightly coloured cast-iron plaque for its clients, and these are today collector's items. The 'fire brigades' were instructed to attend only to fires in houses that carried the company's plaque. In 1843 several of the insurance companies formed a joint Insurance Companies' Fire Brigade.

The Sydney City Council supported a municipal fire brigade from 1844 to 1853. After the failure in 1855 of the first Bill to establish a NSW Fire Brigade, referred to earlier in this section, the Victoria Volunteer Fire Company was formed, it imported its own fire engine, and in 1857 it built the first Australian fire station at 477 Pitt Street South; the building is still standing, listed by the National Trust as a historic building. The volunteer firemen paid a membership fee and ran social functions to raise money, but most of the expenses were met by grants from the insurance companies and businessmen. The volunteer companies played an important part in the days when fire engines could not travel very fast, and indeed might be hauled by manpower, and when many suburban roads lacked paved surfaces. A multiplicity of small fire depots was then more efficient than a central station. A Fire Brigades Bill was eventually passed in 1883, and the Metropolitan Fire Brigades came into existence in 1884.

The development of the Melbourne Fire Brigade was similar to Sydney's, but as the city was founded much later the 'primitive' period was shorter. The initial fire fighting was done with watercarts. After the building of the Yan Yean Reservoir (see p. 155), fire plugs and hydrants were also used, but for many years the piped water supply was inadequate for fire fighting.

Melbourne relied initially on the Insurance Brigade, on a number of municipal fire brigades, and on volunteer brigades. In 1883 the Captain of the Prahran Municipal Brigade called the first of a number of meetings, which resulted in the drafting of a Bill for the creation of a single fire service, which was presented to the Chief Secretary. In 1891 a Metropolitan Fire Brigade Board and a Country Fire Brigades Board were established, and all the individual fire brigades were disbanded.

In Melbourne, as in Sydney, there was discussion on a clause authorising the demolition of buildings to stop a fire, but the clause was retained in the Bill. In 1897 a fire in a city warehouse within a few hours turned into a 'huge conflagration', and the Chief Officer proposed the demolition with explosives of a building to create a fire break. The members of the Board present, who had been called for consultation, opposed the proposal, and shortly afterwards the wind suddenly changed, so that the fire was brought under control without demolition (Maucer 1934, p. 9).

A Head Quarters Fire Station was built at Eastern Hill in 1893; this had a tall tower, from which a large part of the city could be seen, and this aided in the early years in the detection of fires (*Australian Builder* 1953). In 1906 a workshop was built that subsequently supplied much of the equipment for the Victorian Fire Brigades, and also for those in Tasmania and Canberra.

## Fire prevention

Towards the end of the century there was increasing emphasis on prevention. Most fires start small, and it may take several hours before 'flashover' occurs (Figure 71). Until then fires are easily extinguished, but after flashover it becomes a much bigger task. Early notification of a fire is therefore very helpful. In 1885 Sydney installed bell-type fire

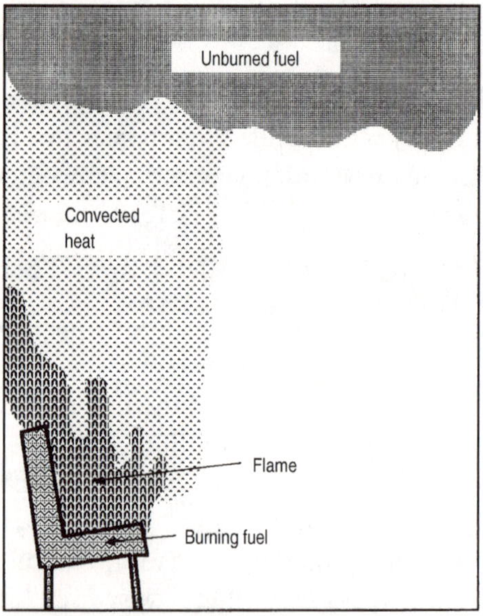


**Figure 71** A fire generates combustible gas that collects near the ceiling. When a small fire spreading slowly along the floor reaches a wall, it spreads upwards on the vertical surface, and when it reaches the ceiling it sets the gas alight, causing a 'flashover'. The entire room is then likely to be filled by flames. (Drawing by Peter R. Smith)

alarms. In January 1890 it became the first city in the world to provide public telephone fire alarms. Thirty, imported from the USA, were fixed on telegraph poles at city corners, the first at the corner of George and Park Streets. The telephones were operable by keys given to police, public bodies and nearby residents, or by breaking a glass panel, and they were connected to a 200-shutter exchange at Fire Headquarters (Adrian 1984, p. 59). In 1898 the Melbourne Fire Brigade secured the telephone connection to its headquarters of many business places.

Automatic extinguishing devices of a kind were already used in the eighteenth century. In 1723 Ambrose Godfrey patented a cask of water to which was attached a cylinder of gunpowder and a fuse. The fire ignited the fuse, which exploded the gunpowder and released the water on the fire. It is reported (AFPA 1974) that this dangerous device extinguished a fire in 1729. In 1806 John Carey devised a system in which open nozzles were attached to empty pipes; the water valve was kept closed by a weight suspended by a combustible cord, which was burnt by a fire. This started a drizzle over the entire area.

This system worked, but the water often did more damage than the fire would have done (Fitzpatrick 1912, p. 112).

The invention of the modern sprinkler is generally credited to Henry S. Parmalee (Figure 72); in his device the water is released by the melting of a fusible link. In Frederick Grinell's sprinkler (1882) the heat-sensitive element is a liquid-filled glass that bursts (AFPA 1974). In both types of sprinkler the water is released at a designated temperature, and it comes from one sprinkler head only if the fire is small, so that the water damage is minimised.

Although sprinklers were an American invention, they are more common in Australia, where building regulations require their use in most commercial and industrial premises, and in multi-storey buildings, to stop small fires from becoming big ones. Most sprinkler installations are today connected to an alarm system, operated by the flow of water, which advises the Fire Brigade that a sprinkler head has opened, and that there is, or at least was, a fire.

Automatic fire alarms have been in use since the 1860s. The original alarms were set off by a rise in air temperature or by radiant heat from a flame. These have now mostly been replaced by smoke alarms, which give an earlier indication, because substantial smoke production occurs before the other indicators become significant. The smoke detectors used since the 1950s contain a small quantity of a radioactive

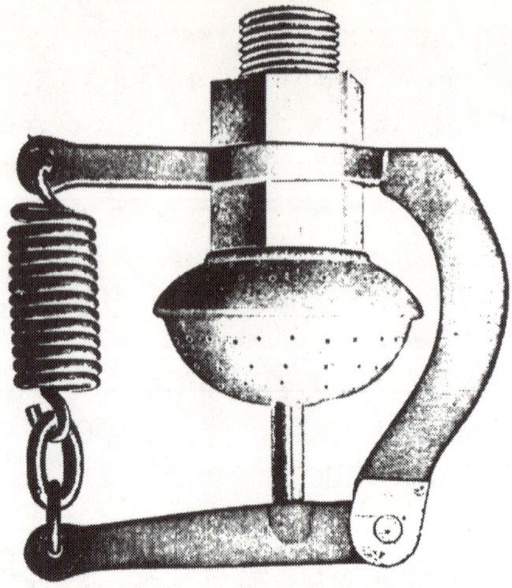

**Figure 72** The Parmalee Sprinkler Mark I, patented in 1874 in the USA, with a fusible link. (Reproduced from Cowan [1978, p. 204])

substance that ionises the air; smoke reduces the current flowing through the ionised air, thereby setting off the alarm.

Many of the world's great cities have suffered serious damage from fire, not merely before the creation of efficient fire-fighting services, but during the twentieth century. The biggest Sydney fire, according to the Fire Brigade's centenary volume (Adrian 1984, p. 193), was in 1890, when most of a city block was destroyed. As mentioned above, a fire of similar magnitude occurred in Melbourne in 1897. However, these are not big fires by international standards.

Three important causes of great fires are war, civil disturbance and earthquakes. In Australia no major fires have resulted from any of the three. In earlier centuries important buildings have been set on fire by lightning, which tends to strike the tallest object. This ceased to be a problem when lightning conductors were invented several decades before the First Fleet sailed; the first lightning conductor was installed in Australia as early as 1790 (see p. 22). While Australia has thus had some advantages, it has established a good record for fire prevention in buildings.

There have been some spectacular fires, however, mostly resulting from wooden construction; for example, the destruction of the first St Mary's Cathedral in 1865 and of the 'Garden Palace' in 1882 (see p. 56).

# 8

# LIFTS AND COMMUNICATIONS

## Building high

Tall buildings were already built by the ancient Romans, and possibly by earlier civilisations. Vitruvius wrote in the first century BC, before Rome became a very big city:

> But with the present importance of the city and the unlimited number of its population, it is necessary to increase the number of dwelling places indefinitely. Consequently, as ground floors could not admit of so great a number living in the city, the nature of the case has made it necessary to find relief by making the buildings high. In these tall piles reared with piers of stone, walls of burnt brick, and partitions of rubble work, and provided with floor after floor, the upper storeys can be partitioned off into rooms to very great advantage. The accommodations within the city walls being thus multiplied as a result of many floors high in the air, the Roman people can easily find excellent places in which to live. (Vitruvius 1960, p. 56.)

In AD 300 a census of Rome listed 1797 *domus* (houses) and 46 602 *insulae* (apartment buildings that occupied whole blocks or 'islands' (Camp 1963, p. 169)). Many people must therefore have lived on upper storeys, accessible only by stairs. We do not know how tall *insulae* were, because where they survive only the lower storeys do, but some Romans were willing to climb stairs to reach the highest seats in the Colosseum, an amphitheatre built in AD 82 that survives; these were 45 metres above ground level, the equivalent of a modern twelve-storey building.

Tall buildings were erected in much smaller cities during the Middle Ages because it was not safe to live outside the city walls. As the population increased, the buildings became taller, and houses of seven storeys survive in a number of cities, such as Edinburgh and Bologna. Before the advent of the passenger lift, the limit was set by the number of stairs people were prepared to climb.

There were thus no structural limitations to the erection of buildings of the order of twelve storeys, even before the invention of the skeleton frame; indeed Culwalla Chambers, the thirteen-storey building (which had lifts) responsible for the imposition of the height limitation in Sydney (see p. 76), had loadbearing walls.

## Vertical transportation

There is no evidence that passenger lifts were used in buildings before the early nineteenth century. They were used in mines, however, probably by the ancient civilisations, certainly in medieval times. Some mines were deep, and the material had to be hoisted to the surface with ropes and pulleys. It would have been a simple matter to use the same gear to take the miners to and from the working face.

The advantages of counterweights were known to the Roman and to many medieval miners, and also the ratchet-and-pawl safety mechanism. However, if the hoisting rope broke, the lift would drop to the bottom. In Roman times lifts were operated entirely by manpower, but in the Middle Ages and the Renaissance animal power and waterwheels were also used.

When factories with steam-power plants were built (see p. 80), it became possible to use them to provide the power for lifts. Steam-powered goods lifts were widely used, and there were a few early passenger lifts. Elliott (1992, p. 328) describes one in an English spinnery in 1835. The power was provided by a pulley driven by a belt from the main engine shaft; the lift was stopped by moving the belt from the drive pulley to an idle pulley. However, not many passenger lifts were installed because of the danger of a serious accident if the hoisting rope broke.

The first safety mechanism was invented by Elisha Graves Otis. It used toothed metal runners at the sides of the shaft, and pawls attached to the lift car. When the car was suspended, the tension of the rope held back the pawls; when the rope was cut, springs pushed the pawls outward to catch the teeth in the side rails, and lock the car in position. Otis demonstrated his safety device at the 1854 Crystal Palace Exhibition in New York by standing on the open platform of a lift and asking an assistant to cut the rope; the platform came to a stop almost immediately. A picture of the test and the admiring audience has been reproduced many times (for example, Elliott 1992, p. 330).

This safety lift was economical in a factory where there already was a steam-power plant, but if a steam engine had to be especially installed for the lift or lifts (Figure 73), the cost was increased.

Elisha Otis's son Charles therefore turned to hydraulic lifts that could be operated off the municipal water supply or by a water tank on the roof, if the pressure was insufficient. The hydraulic ram could

**Figure 73** Steam-powered lift hoisting gear, made in the 1870s. The vertical steam engine is on the right, and its crankshaft is above. The cable drum on the left operates another winding drum on the top floor that hoists the lift car. It works somewhat like a vertical cog railway.
(Otis Elevator Co., Minto, NSW).

operate a hoisting cable, which needed a safety device as in Otis's original proposal, or it could push up a ram that supported the lift (Figure 74), an inherently safer method. The lift was started and stopped by a valve operated by a rope pull (Figure 75).

The second method, which was generally used in Australia, required a hole in the ground deep enough to accommodate the hydraulic ram that pushed the lift up (Figure 74). The first method was preferred in America, because the ram could be laid horizontally in a cellar.

Several other types of lift existed in the nineteenth century. The 'Hydro-Atmospheric Elevator' was used in many early Chicago skyscrapers (see p. 75). When the passenger cage was at ground level, a container in its counterweight (at roof level) was filled with enough water to cause its descent, thus pulling the passenger cage up. The water was then drained at ground level to send the cage down, and the container was refilled at roof level.

The first electric lift was installed in a New York building in 1887 (Parsons 1996); but in the meantime a hydraulic power system was being built in Melbourne.

**Figure 74** Direct-acting hydraulic lift approaching the second storey, pushed up by a ram that requires a casing of equal length in the basement and below ground. The hoisting operation is controlled by a valve in the basement, which is opened and shut by the pull of a rope (see Figure 75).
(Reproduced from Cowan [forthcoming])

**Figure 75** The lift is stopped and started by opening and shutting a hydraulic valve with the pull of a rope, operated by the attendant.
(Otis Elevator Co., Minto, NSW)

## Hydraulic power

The Greeks used hydraulic power (Figure 76), and it was employed extensively from the Middle Ages onwards for flour milling; later it was also the source of power for cutlery manufacture, and in the early textile factories.

When a piped water supply became available, primarily for sanitary purposes, it could also be used as a supply of power, the first source of power that could be transmitted through a conduit, within the limits set by the head of water at the service reservoir. Joseph Bramah, best known as the inventor of the hydraulic press (see also Figure 66(b)), envisaged

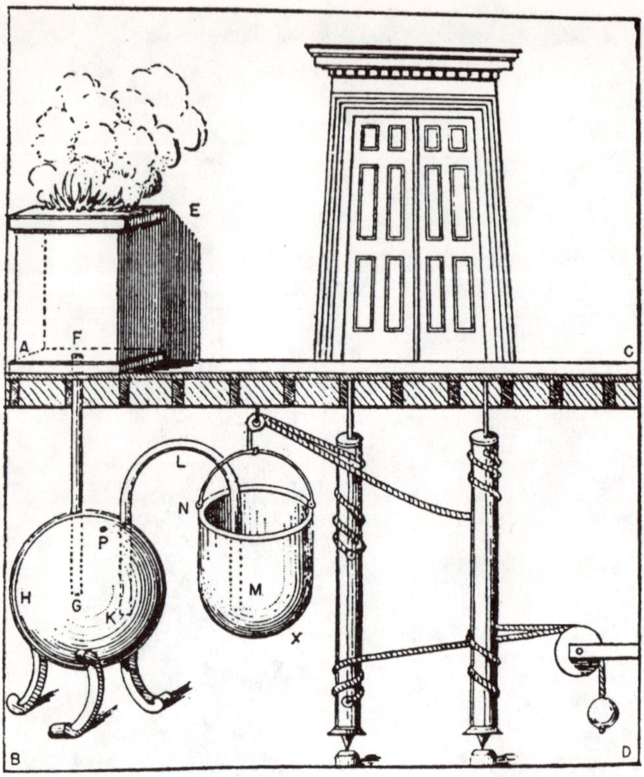


**Figure 76** A hydraulic mechanism for opening the doors of a temple when a fire is lit on the altar for a sacrifice, apparently without the intervention of human hands at the will of a god, described by Hero of Alexandria in the first century AD in his book *Pneumatica* (Book 1, chapter 38).

The heat of the altar fire turns some of the water in the globe (HGK) into steam, and forces some of the remaining water through the siphon into the vessel (M). The increased weight of (M) pulls on the ropes, which revolve the wooden rollers, and cause the doors of the temple to open, revealing the statue of the god. When the fire goes out, the steam condenses, the water from the vessel (M) is sucked back through the siphon, and the counterweight on the right of the picture closes the temple doors.

(Nineteenth-century illustration, reproduced by Cohen & Drabkin [1966, p. 328])

a more potent use in his last patent, registered on 29 April 1812 (McNeil 1972, p. 96), entitled 'Certain imprcvements in the methods of construction, laying down and organizing the main and other pipes for the conveyance of water for the supply of the Metropolis of London etc.' In the specification he explained that the pipes should be connected to workshops and other premises for the operation of machinery. He envisaged a number of pumping stations working onto a high-pressure ring main from which the low-pressure exhaust could be fed to boost the domestic water supply.

Bramah died soon after in 1814, but in the second half of the nineteenth century the use of hydraulic power came close to Bramah's vision. It was a passing phase, because in the twentieth century most of these operations were taken over by electricity; however, it was the first time that power was transmitted over a considerable distance.

The essential invention was the hydraulic accumulator by William Armstrong in Newcastle-upon-Tyne in 1851. This consisted of a tank of water put under pressure by a piston loaded with about a hundred tonnes of rock; it avoided the need for building a tall tower to create the necessary high water pressure. Armstrong's factory supplied the equipment for a variety of heavy-duty applications, such as dockside and railway cranes, dock gates, and lifting bridges. These were connected by pipes to the hydraulic accumulator.

The idea was taken up in Australia in the 1860s by Edward Moriarty, who was then the Engineer-in-Chief, Harbours and Rivers of the NSW Public Works Department, responsible for Hunter River Improvements. He asked Armstrong to design a hydraulic power system for cranes in the Newcastle (NSW) Docks and supply the necessary equipment. This became operational in 1877 (Bairstow 1986, p. 60). The power for the cranes was transmitted through underground pipelines from a powerhouse equipped with boilers, steam engines and hydraulic accumulators. The hydraulic cranes were used until the 1930s, and were then gradually dismantled, today only the empty building of the powerhouse remains.

The success of the Newcastle hydraulic power system led to the building of a public-access hydraulic system in Melbourne, which supplied power for most lifts in the city for many years.

## Hydraulic lifts in Melbourne and Sydney

In the first half of the nineteenth century Australian residential buildings were rarely more than one or two storeys in height. Warehouses and woolstores were built up to five storeys high, and the goods were hoisted up and down on the outside, and taken in through special doors in the facade (see Figure 20).

After the gold rush Melbourne acquired many prestigious commercial buildings, and buildings gradually increased in height. In 1884

Wilhelm Friedrich Prell was proposing to build a four-storey building, which was generally regarded as the limit before lifts were introduced, at 15 Queen Street. During the planning stage he met W. F. Hall, a vice-president of Otis, the American lift manufacturer, who convinced him that he would more than recover the extra cost of erecting a six-storey building with lifts, because the upper floors served by a lift* would yield as much rent as the ground floor. The satisfactory results caused Prell to erect three more buildings in Queen Street with hydraulic lifts, each nine storeys high. Melbourne was then not so far behind Chicago, where in 1891 the term 'skyscraper' was coined for a ten-storey building.

Peter Johns, who migrated to Australia in 1856 and started as an iron fabricator, branched into hydraulic engineering in the late 1870s. In 1877 P. Johns & Co. made and installed their first hydraulic lift, a low-pressure goods lift in a warehouse in Collins Street.

Following a depression in 1878, Melbourne went through a period of speculative building in the 1980s, in which building heights first increased to four and five storeys without lifts, and then to greater heights with lifts. As the number of lifts increased, discussion began on the formation of a public company to supply hydraulic power.

The successful Newcastle (NSW) system had been restricted to one client, as had most previous British systems. However, in 1875 Hull, a city in the north-east of England, had built a small public-access system in 1875 under an Act of Parliament. In 1882 London built a much larger system, followed by Liverpool in 1886. Melbourne became the fourth city in the world to provide a public hydraulic-power supply.

The initiative came from John Coates, a consulting engineer recently migrated from London, who became a Director of Johns & Co., and his twenty-six-year-old nephew, George Swinburne, who soon became a Member of the Legislative Assembly. The Melbourne Hydraulic Power Company was formed, and in 1887 it secured from the Victorian Parliament the right to construct high-pressure mains through the city streets.

A large pumping station was built at a cost of £85 000 at the Australian Wharf in West Melbourne, and 11 km of high-pressure mains

* This was the first passenger lift in Australia, if we do not count a working exhibit by the American Whittier Company, steam-operated and screw-driven, at the Garden Palace Exhibition in 1878 (see p. 56), which was subsequently installed in Toohey's Brewery in Elizabeth Street.

were laid. By the end of 1889 seventy lifts were connected to the system, and the demand for power had increased so rapidly that the company built a new branch line.

The company provided power for wool presses, and for dockside cranes and warehouse whips, but with the increasing height of Melbourne buildings hydraulic lifts became its main business.

In 1883 the American Otis Company appointed an Australian representative in Melbourne, and in 1887 manufacture of hydraulic Otis lifts was started in Melbourne by a local firm, Austral Otis Elevator and Engineering Company. In 1888 the British Waygood Company established the Australian Waygood Elevator Company, also to produce hydraulic lifts in Melbourne. After the Depression of 1891, the Australian Waygood Company combined with Johns to form Johns & Waygood in 1892.

The success of the Melbourne Hydraulic Power Company was used by Coates and Swinburne to persuade a number of investors, initially mostly Melbourne people, to finance a similar system in Sydney. The Act had a smooth passage through the New South Wales Parliament, and the Sydney and Suburban Hydraulic Power Company was floated in 1889. The Minute Book of the early meetings of the Board of Directors is still in existence (Barker 1989, p. 31).

The Water Board did not permit the use of its water supply, which was still inadequate for the growing population, and the company obtained water from a reservoir in Waterloo at the junction of South Dowling Street and O'Dea Avenue (*BEJ* 1891). It is not known whether a hole was dug when the company was formed, or whether there was an existing billabong, part of the chain of lakes that had previously been utilised for the Sydney water supply through Busby's Bore (see p. 153).

The water was pumped from there to the main pumping station, 4 km to the north, at Pier Street, Darling Harbour. The Directors' Report states:

> Your Directors are pleased to report that the erection of the Company's pumping station and of all necessary machinery has been completed in every respect in a very satisfactory manner. Over 8 miles [13 km] of mains have been laid in the city thoroughfares, where consumers will be most numerous . . . 69 machines have already been connected [see Figure 77] to the pressure mains, while contracts have been secured for the working of

37 others, making a total of 107 machines. In addition to the above, enquiries have been received regarding the supply of 67 additional lifts, contracts for many of which are fully expected to be closed on favourable terms. (*BEJ* 1891)

The plant has been described by Tom Dickinson, the company's first engineer and manager (Dickinson 1893–94). The power for the original pump house was provided by a bank of Lancashire-type boilers, with mechanical stokers, which were an innovation at the time. The pumping engines were three high-pressure expansive steam engines of

**Figure 77** Passenger car of the principal lift in the Anthony Hordern Department Store, Haymarket, Sydney (demolished in the 1980s), about 1891. In the era before World War I important commercial buildings treated their lifts as reception rooms. Most had seats, elaborately carved wood panelling, and a mirror.

(Reproduced from Cowan [1997])

the Armstrong-horizontal type. The watertank was constructed from cast-iron plates (Figure 78). The water was pumped into two accumulators, each with a 500 mm diameter ram and a 7 m stroke, loaded with 100 tonnes of gravel, which provided the water pressure.

Every feature in the system, from steam engine to accumulator, was duplicated and could be isolated in case of a break or leak, so that the system could provide a reliable power source for passenger lifts. The drawings were by J. Coates & Co.

Two further pumping sub-stations were built in Woolloomooloo and in Pyrmont in the 1910s. The hydraulic pressure pipes ran to Darling Harbour, where there were many wool presses, Walsh Bay, Circular Quay and the Woolloomooloo Docks, but not to Garden Island, where the naval dockyard had its own hydraulic system. In addition, almost the entire present Central Business District (CBD) was covered. The pipes ran along substantial parts of Sussex, Kent, Clarence, George, Pitt,

**Figure 78** Building incorporating the main pump-house of the Sydney Hydraulic Power Co., now a pub. The water tank made of large, square cast-iron plates with raised circular and diagonal strengthening ridges in the interior, is still on the roof of the lower building. Two of the plates are inscribed 'J. Coates, Engineer, Collins Street, Melbourne. J. Abbott & Co. Ld., Makers, Gateshead, England'. The hydraulic accumulators are still inside the building.

(Photograph by Henry J. Cowan)

Castlereagh, Elizabeth, Phillip and Macquarie Streets, as well as several crossroads. The area between the Haymarket and the present Central Railway Station, however, was not serviced.

In the 1890s and the early twentieth century it was evidently more economical to build a tall building requiring one or more lifts (which at the time meant four to ten storeys), in the part of the town served by the hydraulic mains than anywhere else. Except for suburban developments, the tall buildings of Sydney are still almost entirely in the area served at one time by the hydraulic mains. It is interesting to speculate how far the system influenced the future development of the CBD of Sydney, in much the same way that the Roman street layout influenced the development of the City of London.

## The changeover to electric power

Before reticulated hydraulic power was installed, building owners had the choice between a mechanically operated lift, which might employ screws or hoisting cables, and required a steam or gas engine to provide the necessary power, or a hydraulic lift, which also needed an engine for power. Once a hydraulic power supply became available, the mechanical lifts were at a disadvantage (Selfe 1893–94, p. 42).

Electric lifts suffered initially from the same disadvantage. The first electric lift was installed in Sydney in 1901 before there was an electric distribution system. However, the preference for hydraulic lifts continued for some time after electricity became generally available.

The NSW Department of Labour and Industry reported in 1903 that it had registered 522 hydraulic lifts and 3 electric lifts (Barker 1989, p. 57). In 1919 there were 2369 lifts and whips in the metropolitan area (Barker 1989, p. 54), and most were still operated hydraulically.

The directors of the Hydraulic Power Company had reported satisfactory results until the end of World War I, but the situation changed in the 1920s.

Electric lifts had improved in the interim. The first electric lifts, whether they operated on direct (DC) or alternating (AC) current, used gears to vary their speed. In 1904 Otis built the first gearless lift in America, in which the speed was increased by voltage variation. This allowed much higher lift speeds and faster acceleration, but required a DC supply or rectification of the alternating current. In 1910 the Ward-Leonard control was adopted; each lift had its own motor-generator set, so that the lift could operate on DC while drawing electricity from an AC supply.

In 1923 the American Otis Company in association with the British Waygood Company established a Waygood-Otis factory in Sydney, and in the same year it installed the first electric gearless lift in Culwalla Chambers (Parsons 1996), then the tallest Australian building (see pp. 76 and 170). In addition several other overseas lift manufacturers had arrangements for local assembly.

By the 1920s electric lifts had become much faster than hydraulic lifts, and as reliable. Some people also claimed that they were cheaper. The Standard Electric Company, after converting Norwich Union Chambers in Melbourne from hydraulic to electric operation, published an advertisement that the operating cost had been reduced from £400 per annum to £93 per annum (Barker 1989, p. 109).

The cost of installation became another factor that militated against the choice of hydraulic lifts for new high-rise buildings. Most were of the push-up type, and it was therefore necessary to dig a hole deep enough to accommodate the hydraulic ram. Australia reputedly had the longest hydraulic lift ram in the world, in Melbourne, of 44 metres. The longest in Sydney was in the Hotel Australia, of 32 metres.

In 1920 the Pier Street Pump House was converted from steam to electric operation, and the same was done for the Waterloo pumps in 1926. However, the hydraulic-power system continued in operation for another half-century.

The Sydney Hydraulic Power Company entered electric lift manufacture in 1924, and this soon proved more profitable than the operation of the existing hydraulic system. Hydraulic lifts remained popular for low-rise buildings and for goods lifts, where their slow speed was not a disadvantage, but after 1930 they were generally installed with their own electrically operated pump to make them independent of the piped hydraulic supply.

When the 150-foot height limit was removed in 1957 (see p. 188), many buildings which had been considered 'tall' when they were built in the early years of the twentieth century were demolished to make room for taller buildings. The number of customers of the Sydney hydraulic system declined, but it still had about sixty customers when it was shut down in three stages from June 1974 to June 1975. It had operated for eighty-six years, and outlasted most other hydraulic power systems.

Electric lifts lent themselves better to automatic controls. The rope used to start and stop lifts (Figure 75) was replaced in 1893 by push buttons. Signal control was developed in 1924. It relieved the operator of all duties except opening and closing the doors, and pressing the destination button. The maximum speed was attained automatically, and the lift levelled automatically on arrival. By 1946 a group of cars could be operated automatically as an electronically co-ordinated group, which adjusted itself to the changes in traffic indicated by the calls made and the lift cars taken out of service. In 1950 the first lift system without attendants was installed in an office building in Dallas, Texas (Parsons 1996).

Escalators are almost as old as lifts. The first moving footway was built for the World's Columbian Exposition in Chicago in 1893, and the first moving stair for the Paris Exposition of 1900. Escalators are slower than high-speed lifts, but they can move more people, and they rarely keep them waiting. The period of waiting for a lift is generally short, and if the travel is for more than two floors, the lift will probably get there first; but many people choose the escalator if a lift is not waiting.

## Messengers, tubes and wires

The oldest form of communication is to send a messenger. It was the only method of long-distance communication before the invention of the telegraph. Light signals, flag signals, and semaphores were useful within the range of vision, and homing pigeons could be used over short distances; but until the mid-nineteenth century the only way of getting news in Australia from England or America was to wait for the arrival of a ship that would bring the person or the letter with the information.

Ships had been arriving since 1790; however, the regular mail service to England started only in 1852, and to the American West Coast in 1873 (Australia Post 1988, p. 112). The despatch of the mail was organised to ensure that the place and the latest time when a letter for a particular ship could be posted were known well in advance. The messenger is still useful; we still employ courier services to ensure fast delivery.

Before electrical means of communication became available, there were a number of useful mechanical and acoustic aids. Households that employed servants were often fitted with speaking tubes between the dining room and the kitchen. They were standard equipment in many offices and commercial premises for the purpose of summoning secretaries or giving short instructions. A person speaking into a tube at one end could be heard by a person listening at the other end. Speaking tubes are still in use; the earphones given by most airlines to economy-class passengers consist of empty small-bore plastic tubes fitted with ear plugs. They function almost as well as the electrical earphones issued to business-class passengers.

Pneumatic tubes were widely used in the nineteenth century. They were installed near the ceiling, or buried in the ground. The message was placed in a capsule that was propelled by air pressure (Figure 79). After the start of the telegraph service the major post offices installed pneumatic tubes to the leading newspapers, and to other post offices; the Sydney part of the system was dismantled only in 1985 (Australia Post 1988, p. 70).

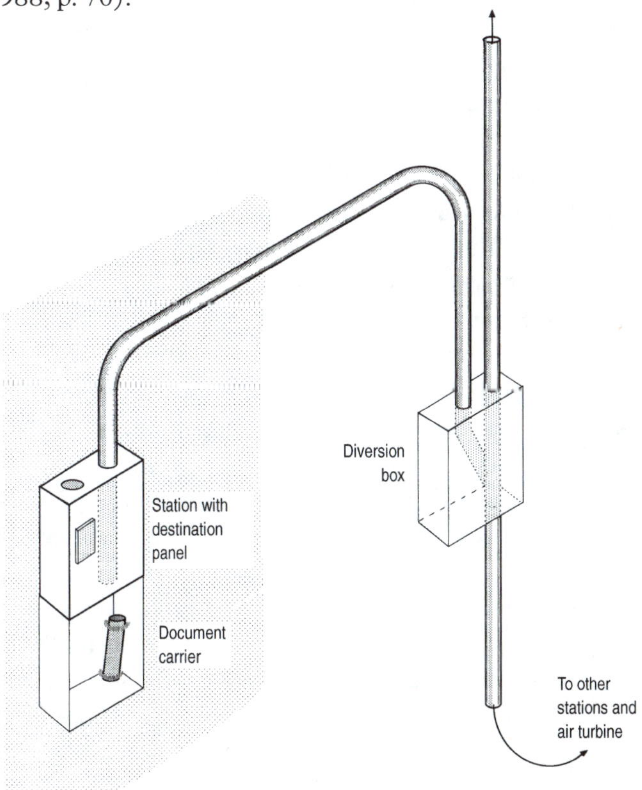


**Figure 79** Pneumatic tube: a capsule containing a document or a small tool or a machine part, is propelled by air pressure.
(Drawing by Peter R. Smith)

Pneumatic tubes were once widely used in commercial premises, sometimes connecting one building to another. Although electronic communications have mostly supplanted them for the transmission of *information*, they continue to be used where it is necessary to transmit real objects, such as documents, cash, or medical specimens, quickly and securely. The new Parliament House in Canberra, completed in 1988, contains about 2 km of pneumatic tubes, while a further 4 km line under the public roads connects it with the Printing Office in another suburb (Joint House Department 1991, pp. 3–36).

Before automatic controls became available, stores preferred to limit the taking of cash to a few trusted people. Some department stores used pneumatic tubes, others installed overhead wires for transmitting money from the salesperson to the cashier. The sales docket and the cash was placed in a capsule, and propelled along the wire by a spring-loaded lever. This saved shoppers the inconvenience of joining a queue to pay, and then going back for their purchase. The system was still in use in some Australian stores in the early 1970s.

The start of the telegraph service in 1854 was a most important innovation. The concept of telegraphy goes back to the mid-eighteenth century when William Watson demonstrated that an electric current could be transmitted through a considerable length of wire, using the earth for the completion of the circuit. In 1831 Joseph Henry constructed an apparatus with a magnetised needle mounted on a pivot in a horizontal position, which could be attracted by a magnet for a shorter or a longer period. However, it was the American artist Samuel Morse's code, in which each letter was represented by a series of dots and dashes (short or long electrical impulses), produced with a hand-operated contact lever (the 'key'), that made telegraphy practicable. The first telegraph companies were formed in America and in England, both in 1845.

The first Australian telegraph line was between Melbourne and Williamstown in 1854. In 1856 Melbourne was connected to Albury, and the following year to Adelaide. The first New South Wales line was a link in January 1858 between the Merchants Exchange in Bridge Street and South Head, replacing a mechanical semaphore system, to notify the arrival of ships. In October 1858 Sydney was linked to Albury, and thereby to Melbourne and Adelaide. The link to Tasmania was established (by undersea cable) in 1860, to Brisbane in 1861, and to Perth in 1877.

The 3000 km overland line from Adelaide to Darwin was completed in 1872, and in the same year an undersea cable was laid from Darwin to Java, which already had a cable connection to London. The first transatlantic cable had been laid in 1866 (Australia Post 1988, pp. 68 and 112).

It now became possible to get a response to a simple message from England, Continental Europe or America in a few hours instead of having to wait several weeks for a reply by sea mail. The telegraph system also greatly improved interstate and intrastate communications. By 1874, when the Sydney telegraph services were relocated into the new General Post Office, New South Wales had 11 000 km of telegraph lines and 180 stations. The Morse machines remained in use for a century; they were replaced by teleprinters only in 1959.

Alexander Graham Bell demonstrated his first working model of a telephone in 1876 in Boston, Massachusetts. It was a variable-resistance transmitter that generated a varying current that produced an acoustic signal at the receiver. He received the patent for the telephone, against several competing claims, in the same year.

The first Australian telephone service was established in Melbourne in 1880, and later that year in Brisbane. Sydney followed in 1882, Hobart and Adelaide in 1883, and Perth in 1887 (Carroll 1888, p. 98). By the end of the 1890s all the capital cities had telephone networks that extended at least 15 km from the centre, and fifty-nine country towns had telephone exchanges.

An important environmental decision was made in 1892, when Sydney's telephone cables were put underground; in most of the world's cities telephone wires were, and often still are, carried on poles above ground. The cables were laid in brick-lined tunnels, about 1.8 metres high, so that one could walk in them. They consumed 7–8 million bricks (Fraser 1989, p. 244), laid in cement imported from England (see p. 81).

The first interstate telephone service between Sydney and Melbourne started in 1907, and it was calculated that the line consumed 300 tonnes of copper wire. For fourteen years it remained a single-line service, which could handle no more than forty calls a day. In 1925 Australia became the first country outside America to introduce the telephone carrier system, whereby additional voice circuits could be superimposed on a single pair of wires using high-frequency separation techniques. After STD (subscriber trunk dialling) had been introduced in 1956, a coaxial cable was laid between Sydney and Melbourne, and placed *underground*.

The Morse telegraph system required a skilled operator who had mastered the code of dots and dashes sent along a single wire. The teleprinter used five parallel wires, transmitting simultaneous on or off signals, and it was operated using a normal typewriter keyboard. Its invention made possible the telex service, which was introduced into Australia in 1954; with it any two subscribers could send messages to one another by teleprinter.

Using five wires, the number of possible combinations was sufficient for languages using the Latin and the Cyrillic (Russian) alphabets, but not for Japanese, which had too many characters. The facsimile machine (FAX), in which letters did not need to be coded, was developed in the late 1970s to overcome this problem, and it was soon found to have much wider application.

# 9

# THE POST WORLD WAR II ERA

## A change of orientation

Australia and its public and commercial buildings changed greatly after World War II; but this did not happen immediately. In the early 1950s most Australians still travelled overseas by boat, partly because air travel was too expensive. The voyage from Sydney to London took a month, so that one expected to stay abroad for several months. Many people took their first overseas trip during their long-service leave, when they were already nearing retirement, or actually after retirement, so that it provided little professional feedback.

In the 1950s many intending travellers still talked about 'Going home', that is, visiting a Britain they had never seen, and which their parents might never have visited, and this was their main, and sometimes their only, destination.

By the end of the 1960s air travel had become much faster and relatively much cheaper. One could now get to London or New York in thirty hours, and shorter and more frequent visits became practicable. Professional people came back with new ideas, and the conquest of the problems created by the distance of Australia from Europe and America, which had started with the completion of the telegraph link in 1872 (see p. 184), was now almost complete, at least for the rapidly increasing number of people who could afford to pay the air fare.

Travelling building professionals no longer talked about 'going home'; they went 'overseas', and many stopped over in Asia, Continental Europe and America. Indeed, America often became the principal destination. Many more innovations in building technology now originated there than in Europe, and America became the major source of technical information that before the war had come mainly from England.

There was, in particular, a renewed interest in tall buildings, which had been banned in Sydney in 1912 (see p. 76), and in other cities soon after.

## Tall commercial buildings

The height restriction had not been a serious restraint in the inter-war years, when a height of 150 feet (46 m) was sufficient for most new buildings. However, after World War II there was increasing pressure for a more intensive utilisation of land in the Central Business District (CBD) of Sydney. The height limit was lifted in 1955, and in 1957 a Heights of Buildings Advisory Committee was set up, to consider all proposals for buildings higher than 150 feet, and most received a sympathetic hearing.

The first building to exceed the former height limit was the AMP Building at Circular Quay, completed in 1961, designed by the architects Peddle Thorpe & Walker and the structural engineers Rankine & Hill. It is a 31-storey, steel-framed building, 117 m high, using long-span steel beams to obtain a column-free interior, welded moment connections for rigid-frame action, steel transfer trusses at plant-room levels, and holes in steel beams for the air-conditioning ducts (Fraser 1989, p. 239).

Many more high-rise buildings were erected in Sydney and in Melbourne in the 1960s and 1970s. New height records were set in 1968 by the 40-storey, concrete-framed Australia Square Tower (170 m high), and then in 1977 by the 60-storey concrete-framed MLC Centre (228 m high) (Figure 80). This remained the tallest building outside North and South America for many years. It was apparently close to the economic limit at the time. The tallest building in Australia at present, the Rialto Tower in Melbourne, also concrete-framed and 60 storeys, is only slightly taller (242 m).

**Figure 80** The Tower Building of the MLC Centre in Sydney under construction in 1973. It is a reinforced concrete tube-in-tube structure (see Figure 83), using precast concrete permanent formwork.

(Photograph by courtesy of Civil & Civic Pty. Ltd.)

In the late 1970s tall buildings were also constructed in Perth and Brisbane. The R&I Tower in Perth, a concrete-framed 52-storey building, completed in 1988, is 214 m high.

Building high greatly increases the expense of the usable floor space (Figure 81). The cost of the columns because of the weight of the building goes up in proportion to height, but the cost of the structure because of wind load increases parabolically. The cost of the building services increases even more rapidly with height, as allowance must be made not merely for the cost of the equipment but also for the loss of rentable space because of the increasing number of lifts required.

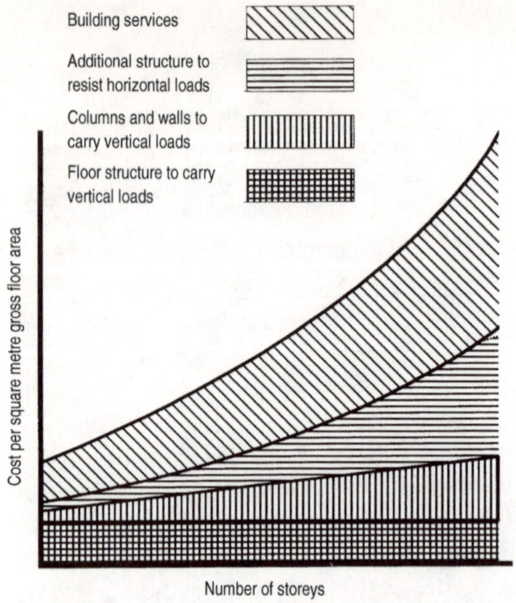


**Figure 81** The cost of building high increases rapidly with height. The floor structure that supports the vertical load, the weight of the building and of the occupants, is the same irrespective of height. The vertical load carried by the columns increases proportionately with height, and so does the corresponding cost. The increased cost of the structure due to wind and earthquake loads increases approximately parabolically with height. The additional cost of the building services increases even faster with height.
(Drawing by Peter R. Smith)

Evidently this increase in cost has to be balanced against the additional income from the greater rentable space created on additional floors of the building, and this depends on the location of the building.

In 1931 the 102-storey Empire State Building in New York (381 m high) set an international record that stood until 1972, when the 110-storey World Trade Center in New York was completed (417 m high). The tallest buildings at present are the twin Petronas Towers in Kuala Lumpur (450 m high). This is a small increase in sixty-five years, set by economic considerations. From a structural point of view it would be feasible to erect buildings at least twice as high (Colaco 1987).

On the other hand, it is nearly double the height of the tallest buildings in Sydney and in Melbourne, where the economic advantage of building high is not as marked as in New York. In Kuala Lumpur the economic height limit may have been stretched by a desire to have the tallest building.

(a)

(b)

**Figure 82** The skyline of Sydney's Central Business District, seen from Mrs Macquarie's Point:
(a) in November 1960; on the right is the AMP Building, the first to exceed the 150-foot limit, nearing completion;
(b) in March 1996; most of the AMP Building on the right is obscured by another tall building.

(Photographs by Henry J. Cowan)

It is likely that even in the Middle Ages and the Renaissance some bishops and princes could not resist the temptation to build a little higher than their rivals; it is certain that this has been a factor in the design of some modern tall buildings. It is not purely a matter of vanity; a person or a company can appreciably increase income by having its name attached to a tall building.

Australian cities developed too late to acquire a noteworthy skyline of spires and/or domes (see p. 54); but during the 1970s the Sydney CBD developed a skyline of tall buildings that could be photographed to great advantage with the Harbour and the Botanical Gardens in the foreground (Figure 82). Melbourne followed soon afterwards, and Perth and Brisbane a little later.

The public reaction has become generally favourable to tall buildings in the CBD, although not to every one of them individually. However, there was opposition to the entire concept of building high in the late 1960s and the early 1970s. When the first Australian international conference on tall buildings, to be held in 1973 at the University of Sydney (Aynsley & Cowan 1973), was being organised, the president of the Melbourne University Conservation Society sent a letter of protest 'against the use of university premises for a conference on tall buildings, which are an environmental disaster on all counts, built by ugly-minded men to get rich quick'.

## The changing structure of tall buildings

Fairly tall buildings were erected without skeleton frames in the Middle Ages. For example the L'Asinelli Tower still standing in Bologna (Cowan 1977, p. 111), built to provide a safe residence during times of civil disturbance, has a height of 98 metres and brick walls about 4 metres thick at the base.

Australian multi-storey buildings in the beginning had load-bearing outer walls of stone or brick, and an inner structure of hardwood posts and beams (see p. 36). From the 1860s on the hardwood was gradually replaced by cast iron, and later by steel and reinforced concrete (see p. 100). However, the Goldsbrough Mort Woolstore at Ultimo was built with load-bearing walls and hardwood posts as recently as 1936. The skeleton frame was originally introduced in Chicago in a 9-storey building in 1885 for constructional, rather than structural reasons (see

p. 75). The first skeleton frame in Australia was built in 1916, by which time the height limit had become law.

During the 1920s and 1930s the advantage of the skeleton frame was mainly constructional. It made it possible to bring materials into the interior of the building, which facilitated their handling. As the outer walls continued to be of conventional brick or stone construction, they could support a substantial load, so that the use of a skeleton frame during the era of the 150-foot height restriction did not result in a great structural economy.

The tallest building in Australia in 1955 was still Culwalla Chambers, built in 1911, 52 metres high, and with load-bearing walls. However, the skeleton frame became a necessity for the taller buildings projected after World War II. Moreover, the new curtain walls (see p. 199) erected by dry construction (that is, jointed with dry sealants instead of wet mortar) could not be utilised without skeleton frames.

The continued use of interior columns was soon questioned. They took up valuable space, and they limited the freedom to rearrange offices. As the height of buildings increased, and the number of lifts increased accordingly, so did the size of the lift core, and fire regulations required that it be encased in concrete. This reinforced concrete *tube* provided a substantial structure that could be utilised for both vertical and horizontal loads.

In a *tube-in-tube structure* the interior columns were eliminated, and the load was carried by an inner tube formed by the service core and an outer tube formed by the columns in the facade (Figure 83; see also Figure 80). As these could be placed partly outside the building, they occupied less valuable space than interior columns. This required quite long spans for the floor structure, and prestressed concrete (see p. 103) was frequently used to reduce its weight.

In every multi-storey building a number of services are required that must be placed above the top storey, such as the hoisting gear for the lifts, and the watertank that provides the gravity head. These projections on the roof looked untidy when seen from a higher level, and as the height of buildings increased, it became customary to place them, together with other services, such as air-conditioning fans, in a service floor at the top of the building. This service floor generally needed to be higher than an ordinary floor to accommodate the equipment, and it did not require any windows. Thus it could be made into a stiff 'top hat' to make a rigid connection between the inner and the outer tubes (Figure 83).

**Figure 83** The concept of the tube-in-tube structure, formed by an inner tube, provided by the reinforced concrete casing of the services core, and an outer tube consisting of the columns and spandrels of the facade. They are stiffened at the top, and sometimes at intermediate levels, by service floors.
(Drawing by Peter R. Smith)

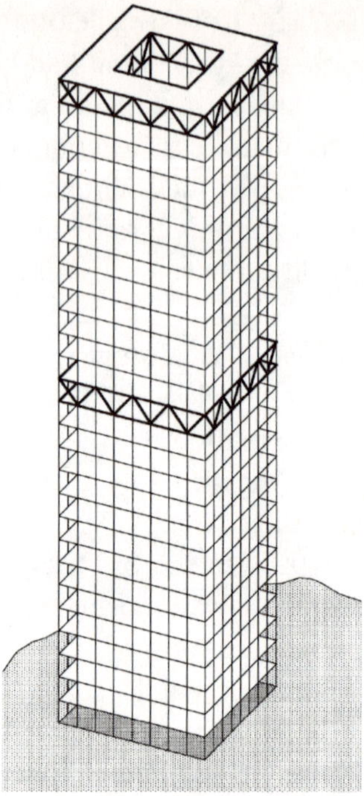

When the construction of tall buildings started in Australia in 1955, it was generally accepted that steel frames would be more economical than reinforced concrete frames for buildings taller than seven storeys; but by the end of the 1960s reinforced concrete frames had generally become cheaper. This Australian trend was part of a worldwide phenomenon: except in the USA, Canada and Japan, where tall steel frames were commonly used, tall reinforced concrete had become more common for high-rise buildings. In Australia this continued until the 1980s, when steel-frame construction made a partial comeback.

## Tall residential buildings

The construction of high-rise *apartment* buildings by the (government-assisted) Housing Commissions of Victoria (see p. 101) and New South Wales in the 1950s and 1960s was part of a worldwide trend at that time. They replaced old terrace houses with inadequate sanitation and very small or no gardens, and the use of high-rise construction provided space for landscaping and playgrounds.

It was a tenet of many theorists of the Modern Movement that slum-dwellers rehoused in a carefully planned community would respond to the new environment. However, this proved unduly optimistic. The public spaces, and particularly the lifts, the long corridors, and landscaped areas hidden from public view were perceived as no man's land, and were often vandalised repeatedly (O'Neill 1974, p. 299).

Charles Jencks attributed the widespread disenchantment with the Modern Movement during the 1970s largely to the failure of its social housing:

> Modern architecture died in St Louis, Missouri, on July 15, 1972, when the infamous Pruitt-Igoe scheme, or rather several of its slab blocks, were given the final *coup de grâce* by dynamite. Previously it had been vandalised, mutilated and defaced by its black inhabitants, and although millions of dollars were pumped back to keep it alive (fixing the broken elevators, repairing smashed windows, repainting), it was finally put out of its misery. (Jencks 1977, p. 9)

The buildings had been designed by a distinguished architect, Minoru Yamasaki, but they were not suitable for housing single working mothers with unruly teenage children, many of whom had guns.

It was only one of a number of subsidised housing estates in America and in England that were ultimately demolished or drastically altered. The Australian experience was less dramatic, but the high-rise public housing schemes were abandoned during the 1970s, because they were unpopular with their inhabitants, and because they created too much opposition from their neighbours and from environmental groups (see also p. 101).

To an increasing extent obsolete working-class houses were refurbished, rather than demolished. They were often gentrified, and this satisfied a new demand for inner-city housing for the middle classes. Planners and conservationists generally considered this conservation preferable to additional high-rise construction.

Tall apartment buildings for the middle and upper classes did not suffer from the problems that had befallen subsidised housing. They met a definite need for single people and for couples who were both working. However, placing high-rise buildings among existing one- and two-storey houses met with understandable opposition from the old residents, and tall residential buildings were increasingly confined to the inner suburbs and the city. Instead there was a growing emphasis on town houses and other forms of medium-density, low-rise accommodation.

## The new interest in heritage buildings

Conservation of old buildings is a relatively recent concept. Before the eighteenth century ruins survived only if the buildings had been solidly constructed of stone, brick or (Roman) concrete. Buildings survived mainly because they continued to serve a useful purpose.

The Parthenon on Athens' Acropolis, built in the fifth century BC, became a church in the seventh century AD. This involved minor structural alterations to the inner columns and the roof, and the building of an apse and some windows. In the fifteenth century it became a mosque, and the only alteration was the building of a minaret. It was ruined only in 1687, when a Venetian warship bombarded a Turkish gun battery on the Acropolis, and the gunpowder stored in the Parthenon exploded. The main reason for its survival for two thousand years was its continued use as a place of worship (Stewart 1965).

The Pantheon in Rome, built in the second century AD, owes its preservation to its conversion into a church in the seventh century. However, the admiration of Pope Urban VIII (Maffeo Barberini) for Roman antiquities did not stop him from taking the bronze girders from its portico in the seventeenth century to serve as the raw material for Bernini's baldachino (Fletcher 1987, p. 872) over St Peter's tomb in St Peter's Basilica, and replacing them with timber.

From the eighteenth century ruins were preserved if they were 'picturesque', and it was quite acceptable to carve one's name on a ruin. In Egypt, where such shallow carvings endure in the dry climate, there are a number of these graffiti. In 1799, when Napoleon Bonaparte's army reached the Temple of Philae, the most southerly point of its incursion into Egypt, the senior officers carved their names on one of the temple walls. In 1885 Kitchener made an inscription on another wall of the same temple on his return from the Sudan Campaign. In 1898 Lt Col Malcolm J. Meade, Her Britannic Majesty's Consul-General, and Mrs Meade carved their names on the interior of the mortuary temple of Ramses III near Luxor.

Apart from isolated earlier instances, serious and widespread restoration of old buildings started only in the nineteenth century. Initially the reconstruction was conjectural, made with good intentions but often without sufficient information, based on guesswork of the original condition. Only rarely were detailed records kept, or photographs or drawings made of the original structure. This conjectural reconstruction of old buildings was criticised by William Morris, and led to his founding of the Society for the Protection of Ancient Buildings in 1877.

The first protective legislation was the (British) *Ancient Monuments Protection Act* of 1882, later extended to historic buildings. The (British) National Trust for Places of Historic Interest or National Beauty was founded in 1895. It was empowered to acquire by gift or purchase buildings to be held in perpetuity for the public. Legislation in France goes back to 1897, and in the USA to 1906.

In Australia there was little interest in conservation before World War II. The National Trust was founded in Sydney by a small group of private citizens in 1945, and was given public status by the *National Trust Act* of the New South Wales Parliament in 1960. Later National Trusts were established in all states. The National Trust can offer advice, which often carries a great deal of weight, but it can protect buildings only by acquisition.

During the 1960s a number of fine old stone buildings were demolished so that the land could be utilised for taller buildings. This loss created support for the growing conservation movement. Books started to appear on Australia's 'historic buildings' (for example, Saunders 1966, National Trust 1969, Roxburgh 1974), and these buildings were included in itineraries of the newly developing tourist trade.

The Australian Heritage Commission was created by the federal parliament in 1975, and there is complementary state legislation in New South Wales, Victoria, South Australia, Queensland and Tasmania.

The National Trust publishes a register of classified buildings and places for each state. Listing by the National Trust does not confer any protection or impose restrictions, but it may strongly influence a decision on an application for replacing an existing building with a new one.

The Heritage Commission publishes a register of the National Estate, that is, of the buildings and places listed by it. This has legal implication for Commonwealth departments, which may range from a prohibition on demolishing a building to restrictions on its development (RAIA 1985, p. 12). The *State Heritage Acts* give legal protection to places under either public or private ownership.

Australia is a member of the International Council on Monuments and Sites (ICOMOS) in Paris, and it supports the International Charter for the Conservation of Monuments and Sites, approved in Venice in 1966, and its amplification adopted by the Fifth General Assembly of ICOMOS in Moscow 1978. These have been adapted to form the

Burra Charter (Burra Charter 1981; Marquis-Kyle & Waller 1992), adopted by Australia ICOMOS at a meeting in Burra, about 150 km north of Adelaide.

The principles of the Burra Charter differ markedly from the conservation practised in Europe and the Middle East in the early years of the twentieth century, and to other concepts of conservation, such as Japan's practice to rebuild certain old shrines at regular intervals:

> *Article 16.* The contribution of all periods of the place must be respected. If a place includes the fabric of different periods, revealing the fabric of one period at the expense of another can only be justified when what is removed is of slight cultural significance and the fabric which is to be revealed is of much greater cultural significance.
>
> *Article 4.* Conservation should make use of all the disciplines which can contribute to the study and safeguarding of the place. Techniques employed should be traditional but in some circumstances they may be modern ones for which a firm scientific basis exists and which have been supported by a body of experience.

The Institution of Engineers, Australia has also taken an active interest in heritage since the 1970s, and it has held biennial heritage conferences since 1982, the most recent (at the time of writing) in 1994 (IPENZ 1994).

It has pinpointed buildings of engineering interest in Sydney (IEAust 1993, pp. 3–30), and it awards Historic Engineering Markers and National Engineering Landmark Plaques to historic engineering works, although only a few *buildings* have been selected so far for recognition.

## New facade materials

The new interest in preserving old buildings has coincided with a drastic change in the use of materials for the facades of buildings. There have been a number of major changes in architectural *style* throughout history, but until the advent of Modern Architecture their effect on architectural *construction* was relatively slight.

The theorists of the Italian Renaissance had some harsh comments on the architecture that is now called Gothic (for example, Filarete 1966, p. 175), as did the theorists of Neo-Gothic four centuries later in describing the 'pestilent art of the Renaissance' (Ruskin 1960, p. 9).

However, the drastic changes in style did not result in a substantial change in construction. The materials for major buildings were still brick and natural stone, carved differently but using the same technology, and wood using the same species of timber, if somewhat different structural principles.

Modern Architecture, however, favoured the use of 'modern materials', notably concrete, metals and glass. In Australia most small houses continued to be built with traditional materials, but the new materials were from the late 1950s used increasingly for the facades of multistorey buildings. This was not, in itself, a technical requirement for high-rise construction. Before World War II tall buildings, far higher than any ever attempted in Australia, were built in America with non-loadbearing facades of natural stone, supported by a skeleton frame. The style changed gradually from decorations in the Gothic or the Renaissance style to Art Deco or almost plain surfaces, but natural stone, generally of known durability, continued to be principal material for the facades.

In Australia there had been a lack of information on the durability of both stone and timber in the early years of the European settlement (see chapters 1 and 2), because there was no record of their performance over a period of time. This had ceased to be a problem by the mid-nineteenth century, provided the process of change in the use of materials and construction technology was sufficiently gradual to allow observation of its effect on durability. In the building boom of the 1960s many designers were in too much of a hurry to use the latest materials and design concepts to allow sufficient time. Concrete is cheaper than natural stone, but the urge to use a 'modern material' was perhaps a more important consideration in its sudden widespread employment on facades. Unlike natural stones, which were laid down by geological action millions of years ago, concrete is produced by the chemical action of water on cement, which continues for several years. The changes in volume that occur cause particular problems if the concrete is exposed on the facade of a building. Deterioration occurs especially with off-the-form concrete (Figure 84), because the chemically active material, the cement, covers the entire visible surface.

One solution widely used in the 1960s was the use of exposed aggregate, with a minimum of visible cement, on precast concrete panels, joined with dry sealants, not with wet mortar; but the size of the aggregate created a 'readability' problem. If small aggregate (about 10 mm) in different colours was selected, it looked good if viewed at close quarters;

however, at a great distance the different-coloured aggregates merged into an unattractive grey or brown.

To overcome this problem, broken concrete surfaces were introduced in the 1970s, produced by splitting concrete blocks, by hammering ribs to break off pieces of concrete, or by casting ropes or bars close to the surface and breaking off the concrete surfaces when they were pulled out (illustrations of these surface finishes are reproduced in *Constructional Review* 1977, pp. 55, 59 and 62). These rough surfaces 'to emphasise the strength of the concrete' were acceptable at a distance; but at close quarters the broken concrete surfaces looked like broken concrete, inappropriate for a prestigious building.

In the 1980s there was a return to more traditional facade materials; in particular natural stone came back into use, as a dry-jointed veneer attached to the building in the final stages of construction. However, all

**Figure 84** Dark irregular streaks on an off-the-form concrete wall, caused by rainwater, photographed in 1986, three years after completion. The problem was aggravated by the lack of a coping to throw the water off the concrete surface, customary in Traditional but not in Modern Architecture.

(Photograph of part of the Architecture School of the University of Sydney by Henry J. Cowan; many photographs showing similar problems could be taken all over Australia, and indeed all over the world)

the larger cities were left with a number of concrete facades whose surface finish compares unfavourably with that of the stone and brick buildings of the era before World War II.

The use of metals for curtain walls also created some problems. Metals are produced from their ores by the expenditure of energy, and when they oxidise they release energy, so that the process is self-perpetuating as long as there is an adequate supply of oxygen. Durability depends on the formation or artificial application of a skin that shuts out the supply of oxygen.

The metal favoured for curtain walls in Australia in the 1960s and 1970s was aluminium. During World War II there had been an enormous increase in its production for military aircraft, and it became an important building material after the war.

Aluminium oxidises rapidly and forms a hard and coherent skin, which can be thickened by anodising, an electrolytic process. However, this protective skin alters with time. Old aluminium-curtain walls, such as that on the Alcoa Building in Pittsburgh built in 1953, show signs of deterioration.

The same applies to stainless steel, which forms a very thin hard film of chromium-rich oxide. Stainless steel at the Sydney Opera House, admittedly a building located in a salt-laden atmosphere, corroded severely in a little over twenty years.

Durability is not a problem with glass, probably the most durable of all building materials, if one does not throw bricks at it, or explode bombs nearby.

Glass was one of the favoured materials of Modern Architecture overseas, because of the crystalline appearance and the geometric simplicity of glass walls (Pevsner & Richards 1973, pp. 187–92; Wingler 1969, p. 31), and because it gave a clear visual demonstration of the lightness of the skeleton frame with curtain walls, as opposed to the traditional structure with load-bearing walls. In Australia it was treated with caution in the immediate post World War II era, because of the heat transmitted through ordinary window glass by solar radiation.

The first *heat-absorbing* glass was developed in the 1940s, but as it was green it gave an unpleasant colour to the daylight transmitted. Bronze glass was introduced in 1965, and grey glass in 1966 (Calderone 1988).

The first *heat-reflecting* glasses, using a chromium metal coating, became available in Australia in 1972, and in 1979 coatings of titanium nitride and stainless steel (Calderone 1988). The glass is coated with a metallic film, which reflects some of the radiation back to the exterior of the building. The thermal efficiency of windows is greatly increased by the use of double glazing, using a plastic insert to break any thermal path through the metal window frame; the heat transmission is reduced by the low thermal conductivity of the sealed air space.

Heat-reflecting glass-curtain walls thus became more common in the late 1970s and the 1980s, even in Queensland. They do not have a durability problem, but they have an impersonal quality. This is not a serious defect as long as they reflect interesting buildings around them. One would not wish to be surrounded entirely by buildings faced with reflecting glass walls reflecting one another. Moreover, the reflected heat and light generated by the glass facades may create serious environmental problems for neighbouring buildings, and for passers-by, particularly for drivers of motor cars.

The buildings with walls of concrete, metal and glass have not received the same measure of public approval as those faced with more traditional materials. It remains to be seen whether there will be the same support for listing them as Heritage Buildings and encourage their preservation for future generations.

## The increased importance of building services

More far-reaching than the change in facade materials, although less visually evident, has been the great increase in the complexity and cost of the building services for major buildings.

Except during the Roman Empire, building services were minimal until the eighteenth century. Villard de Honnecourt, a Gothic master builder, described a device for keeping the bishop's hands warm during mass in an unheated, and at that time unheatable, cathedral:

> If you wish to make a hand warmer, you must first make a kind of brass apple with two fitting halves. Inside this brass apple there must be six brass rings, each with two pivots, and in the middle there must be a little brazier with two pivots. The pivots must be alternate in such a way that the brazier always remains upright, for each ring bears the pivots of the others. If you follow the instructions and the drawings, the coals will never drop out, no matter which way the brazier is turned. A bishop may freely use the device at High Mass; his hands will not get cold as long as the fire lasts. (Villard 1966, p. 66)

In medieval Britain the only building services were heating by open fireplaces and lighting by candles and rushes. The same applied to New South Wales for several decades after the first settlement, except that whale-oil lamps were used from an early date.

Piped coal gas and water became available from the 1840s on, and piped sewerage from the 1850s, but the completion of the sewerage systems took many decades (see chapters 5, 6, and 7). The reticulation of electricity and telephones started in the 1880s. These services were gradually made available to suburbs and country areas. With the addition of a water heater, they were still the services installed in buildings at the end of World War II, and they did not greatly affect their design. Air conditioning was installed in some rooms and a few complete buildings from the 1930s (see p. 142).

The scope of the services increased shortly after the war. In the 1960s air conditioning was installed in many of the new tall buildings, and it soon came to be regarded as an essential service for office buildings.

In the meantime fluorescent tubes had been introduced from America, and their standard lengths provided the module for the 'integrated ceilings', suspended from the reinforced concrete slab of the floor above. The space between the floor slab above and the ceiling accommodated the light fittings, the electric wiring, and any air-conditioning ducts, all of which were thus easily accessible. Lighting levels became higher and electric light was increasingly substituted for daylight. This produced a substantial heat load, which was removed by air conditioning.

During the post-war era there was also increased use of modular non-load-bearing partitions in lieu of load-bearing partitions. These could be adjusted to the changing requirements of the tenant, or of a new tenant, giving an increased flexibility to the office layout.

The electrification of office equipment started in the 1950s, when most typewriters and calculators were still manually operated, and thus required no services except good lighting. Letters and books could then only be copied by hand, by retyping, or with a photographic camera. For multiple copies stencils had to be cut by typing or by handwriting. During the 1960s electric typewriters and calculators became more common, and the first copying machines came into use.

The first Australian electronic digital computer was installed by the Physics School at the University of Sydney in 1956. By the 1960s the

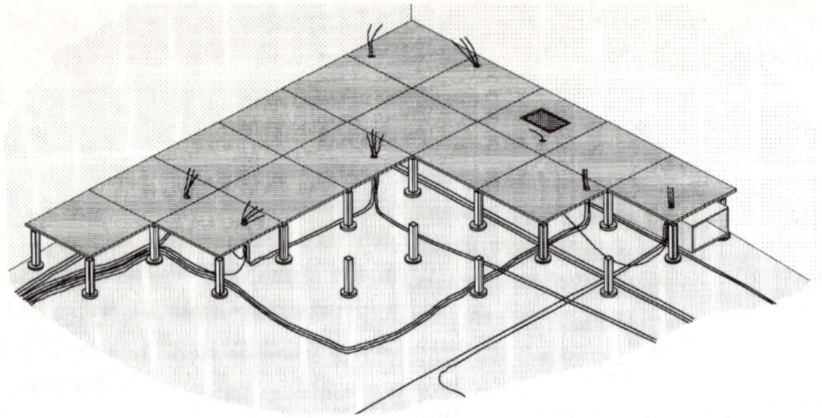

**Figure 85** Raised floor to supplement or replace the suspended ceiling. It is particularly useful for accommodating computer cables, and it can serve as an air-conditioning plenum. If the floor is installed in an existing building, the doors can easily be altered to allow for the raised floor level, but this is not practical for the lifts; ramps are required from the old floor level retained in the lift lobby to the new floor level.

(Drawing by Peter R. Smith)

potential of computers for the technical aspects of building design and for commercial operations had become evident, and in the 1970s computers became a part of standard office equipment. Typewriters, calculating machines and slide rules were more and more replaced by computer terminals and printers. The heat generated by them was removed by additional air conditioning.

This new equipment was generally arranged in flexible workstations, and it was more convenient to bring the numerous cable connections up from the floor than down from a suspended ceiling. Installation of a raised floor (Figure 85) in an existing building posed some problems, but these do not arise if a building is initially designed with one.

By 1980 the building services had become much more expensive than the structure of a tall building, and the provision of the most modern services had become a major factor in determining the rental value of office space. Thus the useful life of an office building now depended largely on the ability to update the services economically. This is a marked change from the concept that important buildings should be designed to last for ever.

# GLOSSARY

**Note:** *Italics* indicate a cross-reference to another term in the glossary.

**absolute humidity** The mass of water vapour in a unit mass of air; it is not dependent on temperature. The *relative humidity* depends also on the temperature.

**AC** Alternating current.

**adze** A traditional tool for working timber, now rarely used. It has a cutting blade mounted at right angles to a long wooden shaft.

**aqueduct** An artificial channel for the conveyance of water or sewage. Most aqueducts are underground, and thus invisible; hence the term is often applied particularly to the visible part of the system carried on bridges.

**arcade** Originally a range of arches supported by columns, frequently supporting a passage sheltering pedestrians from the rain and the sun. Hence a covered passage with shops on both sides.

**ASHRAE** American Society of Heating, Refrigerating, and Air-Conditioning Engineers.

**beam** A horizontal member of the structure of the building, made of timber, iron, steel, or reinforced concrete.

**BEJ** Building and Engineering Journal.

**bluestone** A dark bluish rock. In Melbourne the term is generally used for basalt, but in Adelaide it is more often used for slate.

**box girder** A hollow *beam*, built up from pieces of timber, iron or steel. It usually has greater depth than a solid beam made of the same weight of material, and can therefore carry more load.

**cast iron** Iron containing 2–4 per cent of carbon. It has a lower melting point than *wrought iron*.

**cavity wall** A wall, usually of brick, built of an inner and an outer leaf. The space between them, about 50 mm, drains any water penetrating the outer leaf.

**CBD** Central Business District.

**CIE** Commission Internationale de l'Éclairage.

**curtain wall** A thin wall supported by the *skeleton frame* of the building, as opposed to a *load-bearing wall*.

**damp course** (Incorrect) abbreviation of *damp-proof course*.

**damp-proof course** Impervious horizontal layer inserted into a pervious wall, such as a brick wall, to prevent the upward passage of moisture.

**DC** Direct current.

**dry-bulb temperature** The temperature indicated by a normal thermometer, as distinct from the wet-bulb temperature, which is used to determine the *relative humidity*.

**dry construction** Building without the use of wet mortar, concrete, or plaster.

**elevator** The same as a *lift*.

**encaustic tile** A tile whose coloured decoration has been fixed by the application of heat. It was used in medieval buildings, and the technology was revived in the Victorian era, initially for Neo-Gothic buildings.

**flat plate** A concrete slab supported directly on the columns without column capitals and beams.

**flat slab** A concrete slab supported directly on the columns without beams, but with column capitals.

**fluorescent lamp** A lamp consisting of a glass tube coated inside with a phosphor that emits visible light when struck by (invisible) *ultraviolet* rays, produced by an electric discharge between the two electrodes in the tube.

**freestone** A building stone that can be freely cut in any direction.

**geometric staircase** A staircase formed by cantilevered stone treads built into the wall, which support one another by shell action.

**girder** A large *beam*.

**header** A brick or stone laid across a wall, so that the shorter length, or head, is visible on the face of the wall.

**heat pump** An air-conditioning installation that can act either as a heating or a cooling unit.

**HRA** Historical Records of Australia.

**HRNSW** Historical Records of New South Wales.

**HUD** Department of Housing and Urban Development, Washington, DC.

**hydrant** A connection to a water main, usually for the purpose of extinguishing fires.

**hydraulic lime, cement** A lime or cement that sets and hardens under water, and is then water-resistant. Ordinary lime is washed out if submerged in water for a long period, because hydrated lime is soluble, but there are a few lime deposits from which a hydraulic lime can be obtained. All materials today called cement are 'hydraulic', that is they are water resistant; but in the nineteenth and early twentieth centuries the term 'cement' was used for some materials that lacked this hydraulic property.

**ICE** (British) Institution of Civil Engineers.

**IEAust** Institution of Engineers, Australia.

**IPENZ** Institution of Professional Engineers, New Zealand.

**joist** A small *beam*.

**laitance** A layer containing an excess of cement and very fine sand, brought to the surface of the concrete if the mix contains an excess of water. It has poor durability and strength.

**lead crystal** Glass made with a flux of lead monoxide; it has a high refractive index, and consequently a desirable sparkle and brilliance.

**lift** An enclosed platform for conveying passengers or goods in a multi-storey building. The same as an elevator.

**load-bearing wall** A wall that supports part of the weight of the building, as opposed to a *curtain wall*.

**loam** Clay containing a substantial admixture of silt and sand.

**natural cement** Cement made from a naturally occurring mixture of limestone and clay, as opposed to *Portland cement*. The first commercially produced cements, both in England and in Australia, were natural cements; but production ceased when Portland cement became freely available.

**NEAC** National Energy Advisory Committee, Canberra.

**Otto cycle** The working cycle of a four-stroke heat engine. The cycle of suction, compression, explosion at constant volume, expansion, and exhaust occupies two revolutions of the crankshaft.

**pediment** A low-pitched gable above a portico used in classical architecture.

**pig iron** The term originated from the traditional method of producing *cast iron*: running the liquid metal into a channel in a bed of sand, and from there into smaller channels. The main channel was called a sow, and the smaller channels the pigs. Hence pig iron was the primary form of metallic iron, the raw material for subsequent manufacture.

**plenum** A space in a ceiling or floor used for artificial ventilation or air conditioning. It is generally kept at a pressure slightly above atmospheric for air supply, and slightly below atmospheric for air exhaust.

**plenum duct** A duct acting as a *plenum*.

**plinth** The projecting base of a wall.

**polychrome brickwork** Brickwork made from bricks of various colours, arranged in patterns.

**Portland cement** Cement made from limestone and clay or shale dug from two different locations, as opposed to *natural cement*, which is made from a natural admixture of both materials. The term 'cement' today normally means Portland cement.

**prestressed concrete** Concrete which is precompressed in the zone where tensile stresses occur under load; consequently cracking in the concrete due to tension is prevented. This is usually done by tensioning steel with hydraulic jacks, and then anchoring it to the ends of the concrete beams or slabs. Because there are no cracks, it is possible to use much higher stresses in the steel, and hence also in the concrete. As a result a prestressed concrete floor structure weighs less than one of normal reinforced concrete.

**quoin** An outer corner of a wall. Hence dressed natural stone placed at the corners of a brick building, or face brick placed at the corners of a building with stone rubble walls. The quoins may be for additional strength, or for decoration, or both.

**RAIA** Royal Australian Institute of Architects.

**reciprocating movement** In an engine, movement alternately backward and forward.

**relative humidity** The ratio, expressed as a percentage, of the water vapour actually present in the air (the *absolute humidity*) to that which would be present at the same temperature in water-saturated air. The relative humidity of water-saturated air is 100 per cent.

**reverberatory furnace** A furnace in which iron is melted on a shallow hearth by a flame passing over the iron and heating a low roof. Thus it does not absorb any carbon from a coke fire.

**RIBA** Royal Institute of British Architects.

**semaphore** A signalling apparatus, generally consisting of a vertical post and one or more movable arms, which have lanterns at night.

**shale** A sedimentary rock formed mainly by clay and silt particles.

**skeleton frame** A frame that carries the loads, while providing only a very small part of the envelope of the building; as opposed to *load-bearing walls*, which perform both functions. However, load-bearing walls must be built as the building rises to support the floors, whereas curtain walls supported by a skeleton frame can be erected at a late stage in the construction.

**slaked lime** Calcium hydroxide, made by slaking, that is adding water, to quicklime (calcium oxide). The old-fashioned product was made in a lime maturing pit, and the 'lime putty' required some weeks to mature. It is now common practice to use for lime mortar 'dry hydrate', a dry powder made in a factory and supplied in paper bags.

**SMH** Sydney Morning Herald.

**stanchion** A large column.

**steel** An alloy of iron and carbon, with a carbon content between 0.1 and 1.7 per cent. Iron with less carbon is called *wrought iron*, and iron with more carbon is *cast iron*. Before the invention of the Bessemer process in the mid-nineteenth century steel was a very expensive material. By the 1920s both wrought iron and cast iron had been superseded by steel for structural purposes.

**storey post** A wooden column used in 'colonial' multi-storey construction.

**strain** Change of length per unit length.

**stress** Force per unit area.

**stretcher** A brick or stone laid with its longer side parallel to the wall.

**teleprinter** A telegraphic instrument that enables messages to be transmitted by typing, instead of using a Morse key.

**trachyte** A fine-grained igneous rock.

**Triassic sandstone** A sandstone laid down during the Triassic period, about 230 million years ago. The Triassic period is older than the Jurassic, but more recent than the Permian.

**ultraviolet** Radiation with a wavelength shorter than violet, the shortest of visible light (390 nm). Hence it is invisible to the human eye.

**Ward-Leonard control** A method of speed control for large *DC* electric motors, employing a variable-voltage generator to supply the motor, which in turn is driven by an *AC* motor.

**wattle-and-daub construction** Construction of coarse basket work of twigs woven between upright poles, plastered with mud.

**Welsbach mantle** An incandescent mantle composed of cotton impregnated with thorium and cerium oxides, invented by Carl Auer von Welsbach in 1885 to improve the performance of gaslight, but later also used for other lamps.

**whip** Hoisting gear for small loads, generally fixed to the outside of a building.

**wrought iron** Iron containing less than 0.1 per cent of carbon. It has a higher melting point than *cast iron*.

# BIBLIOGRAPHY


Abrams, Duff A. 1919, *Design of Concrete Mixes*, Bulletin no. 1, Structural Research, Lewis Institute, Chicago, 20 pp.

Adrian, Colin 1984, *Fighting Fire!*, Allen & Unwin, Sydney, 271 pp.

AFPA (Australian Fire Protection Association) 1974, 'Centenary of the automatic sprinkler system.' *AFPA Fire Journal*, vol. 1, no. 1, Sept., p. 31.

Ahern, W. R. 1979, 'Fifty years of comfort air conditioning', *Australian Refrigeration, Air Conditioning and Heating*, Jan., pp. 17–19.

Aird, W. V. 1961, *The Water Supply, Sewerage, and Drainage of Sydney*, Metropolitan Water, Sewerage and Drainage Board, Sydney, 347 pp.

Alberti, Leon Battista 1955 (1755), *Ten Books of Architecture*, (trans. J. Leoni), Tiranti, London, 256 pp. (facsimile of first English edn).

Anderson, Gordon F. 1955. *50 Years of Electricity Supply*, Sydney County Council, Sydney, 257 pp.

ASHRAE 1975, *Energy Conservation in New Building Design*, ASHRAE Standard 90–75, ASHRAE, New York, 53 pp.

—— 1977, *ASHRAE Handbook of Fundamentals*, ASHRAE, New York, pp. 8.19–8.28.

—— 1989–92, *ASHRAE Handbook*, American Society of Heating, Refrigerating and Air-Conditioning Engineers, Atlanta: *Fundamentals*, 1989, approx. 750 pp.; *HVAC Applications*, 1991, approx. 750 pp.; *HVAC Systems and Equipment*, 1992, approx. 750 pp.

Atkinson, E. 1875, *Physics—Experimental and Applied, Translated from Ganot's Éleménts de Physique*, 7th edn, Longmans, Green & Co., London, 908 pp.

Australia Post 1988, *The City's Centrepiece: The History of the Sydney G.P.O.*, Hale & Iremonger, Sydney, 114 pp.

*Australian Builder* 1953, 'The Eastern Hill Fire Station', *Australian Builder*, June, pp. 372, 373 and 426.

*Australian Dictionary of Biography*, vol. 2, 1788–1850, I–Z 1967, Melbourne University Press, Melbourne, p. 584.

Aynsley, R. M. & Cowan, H. J. 1974, *Proceedings of the Australian and New Zealand Conference on the Planning and Design of Tall Buildings, 14–17 August, 1973*, University of Sydney, Sydney, 683 pp.

Bairstow, Damaris 1986, 'Hydraulic power and coal loading at Newcastle Harbour, New South Wales', *Australian Journal of Historical Archaeology*, vol. 4, pp. 57–66.

Baker, H. 1961, *Historic Buildings: Parramatta*, Cumberland County Council, Sydney, 35 pp.

—— 1969, *Elizabeth Bay House*, Angus & Robertson, Sydney, 74 pp.

Baker, R. T. 1915, *Building and Ornamental Stones in Australia*, published by authority of the Government of New South Wales, Technological Museum, Sydney, 169 pp.

Balint, Emery 1984, *Historic Records of Sydney City Buildings*, Report, School of Building, University of New South Wales, Sydney, 405 pp.

——, Howells, Trevor & Smyth, Victoria 1982, *Warehouses and Wool Stores of Victorian Sydney*, Oxford University Press, Melbourne, 162 pp.

Banham, Reyner 1969, *The Architecture of the Well-tempered Environment*, Architectural Press, London, 295 pp.

Barker, Ray M. 1989, Vertical Transport in Sydney c. 1880–1915: The era of dominance of hydraulic lifts in the Central Business District, MA thesis, University of Sydney, Sydney, 114 pp., 8 appendices.

Bassett, Jan 1993, *The Oxford Illustrated Dictionary of Australian History*, Oxford University Press, Melbourne, 304 pp.

Bedford, Thomas 1964, *Basic Principles of Ventilation and Heating*, 2nd edn, H. K. Lewis, London, 438 pp.

*BEJ* 1891, *Building and Engineering Journal*, 15 July; quoted in *Highlights from Our History, 1889–1989*, EPL-Kone Elevators, Sydney, 1989, p. 1.

Billington, Neville S. & Roberts, Brian M. 1982, *Building Services Engineering: A Review of its Development*, Pergamon Press, London, 537 pp.

Birmingham, Judy, Jack, Ian & Jeans, Dennis 1979, *Australian Pioneer Technology: Sites and Relics*, Heinemann, Melbourne, 200 pp.

Blainey, Geoffrey 1956, *One Hundred Years of Johns & Waygood Limited 1856–1956*, Johns & Waygood, Melbourne, 72 pp.

Blakey, F. A. 1965, 'Towards an Australian structural form—the flat plate', *Constructional Review*, vol. 54, Sept., pp. 115–17.

Booker, Stephen Ronald 1979, George McRae: City architect and building surveyor, BArch thesis, University of New South Wales, Sydney.

Bradley, William 1969, *A Voyage to New South Wales: The Journal of Lieutenant William Bradley RN of HMS* Sirius *1786–1792*, vol. I, facsimile, Public Library of New South Wales, in assoc. with Ure Smith, Sydney, 495 pp.

Brady, Frank 1994, Electricity Supply in NSW: The First Century and Beyond, paper delivered to the Sydney Division of the Institution of Engineers, Australia, 17 Oct., 12 pp.

—— 1996, personal communication, March.

Bridges, Peter 1995, *Foundations of Identity: Building Early Sydney 1788–1822*, Hale & Iremonger, Sydney, 186 pp.

Broadbent, James 1992, 'Building in the colony', in *The Age of Macquarie*, eds James Broadbent & Joy Hughes, Melbourne University Press, Melbourne, pp. 157–171.

Brown, Geo. 1937, *Victorian Building Stones*, Mines Department, Melbourne, 23 pp.

Brown, Joyce M. 1966–67, 'W. B. Wilkinson (1819–1902) and his place in the history of reinforced concrete', *Transactions of the Newcomen Society*, vol. 39, pp. 129–42.

Brown, Shylie & Brown, Ken 1995, *Parramatta: A Town Caught in Time*, Hale & Iremonger, Sydney, 138 pp.

Bures, Susan 1987, *The House of Wunderlich*, Kangaroo Press, Sydney, 160 pp.

Burra Charter 1981, *The Australia ICOMOS Charter for the Conservation of Places of Cultural Significance (The Burra Charter)* 1981, Australia ICOMOS, n.p., 12 pp.

Byrne, Bill (Conrad & Gargett, Architects) 1995, personal communication, Dec.

Calderone, Ignatius, (Pilkington ACI, Architectural Glass Division) 1988, personal communication, 19 Feb.

Camp, Le Sprague de 1963, *The Ancient Engineers*, Rigby, Adelaide, 408 pp.

Cannon, Michael 1993, *Melbourne after the Gold Rush*, Loch Haven Books, Melbourne, 473 pp.

Capon, Joanna 1991, *Plaster Work: A Practical Guide*, Random House Australia, Sydney, 94 pp.

Carrier Air Conditioning Company 1965, *Handbook of Air Conditioning Design*, McGraw-Hill, New York, approx. 750 pp.

*Carrier News* 1982, Golden Anniversary issue, 12 pp.

Carroll, Brian 1988, *The Engineers: 200 Years at Work in Australia*, Institution of Engineers, Australia, Canberra, 311 pp.

Casey, Maie et al. 1963, *Early Melbourne Architecture 1840–1888: A Photographic Record*, Oxford University Press, Melbourne, 184 pp.

Chadwick, Edwin 1965 (1842), *Report on the Sanitary Condition of the Labouring Population of Great Britain*, abridged and ed. M. W. Flinn, Edinburgh University Press, Edinburgh, 443 pp.

Chapman, Don 1986, *1788: The People of the First Fleet*, Doubleday, Sydney, 191 pp.

CIE 1970, *Daylight: International Recommendations for the Calculation of Natural Daylight* publication no. 16 (E-3.2), Commission Internationale de l'Éclairage, Paris, 29 pp.

Clark, Ralph 1981, *The Journal and Letters of Lt. Ralph Clark 1787–1791*, eds Paul Fidion & R. J. Ryan, Australian Documents Library, in assoc. with Library of Australian History, Sydney, 344 pp.

Clarke, Robert 1965, 'Whale fisheries', *Encyclopedia Britannica*, Chicago, vol. 23, p. 555.

Clifton-Taylor, Alec 1962, *The Pattern of English Building*, B. T. Batsford, London, 384 pp.

Cohen, M. R. & Drabkin, I. E. 1966. *A Source Book of Greek Science*, Harvard University Press, Cambridge, Mass., 581 pp.

Colaco, Joseph P. 1987, 'Structural concepts for ultra-tall buildings', *IABSE Symposium, Versailles: Concrete Structures of the Future*, International Association for Bridge and Structural Engineering, Zurich, pp. 533–8.

Collins, A. R. (ed.) 1983, *Structural Engineering: Two Centuries of British Achievement*, Tarot Print for the Institution of Structural Engineers, London, 200 pp.

*Constructional Review* 1977, anniversary number, vol. 50, Nov., pp. 9–77.

Cowan, Henry J. 1977, *The Master Builders: A History of Structural and Environmental Design from Ancient Egypt to the Nineteenth Century*, John Wiley, New York, 299 pp.

—— 1978, *Science and Building: Structural and Environmental Design in the 19th and 20th Centuries*, John Wiley, New York, 374 pp.

—— (ed.) 1980, *Solar Energy Applications in the Design of Tall Buildings*, Applied Science Publishers, London, 333 pp.

—— 1981, 'Domes of the transition period between the classical era and the age of engineered structures', *Architectural Science Review*, vol. 24, pp. 1–4.

—— (ed.) 1983, *Predictive Methods for the Energy Conserving Design of Buildings*, Pergamon Press, Sydney, 118 pp.

—— (ed.) 1984, *Energy Conservation in the Design of Multi-storey Buildings*, Pergamon Press, Sydney, 249 pp.

—— 1993, 'The scientific and political disputes which preceded the development of the present theory of structures', *Transactions of the Institution of Engineers, Australia—Multi-disciplinary Engineering*, vol. GE17, no. 1, pp. 51–8.

—— 1994, 'A history of structural concrete and its adoption in Australia as a major building material', *Transactions of the Institution of Engineers, Australia—Multi-disciplinary Engineering*, vol. GE18, no. 2, pp. 63–74.

—— (forthcoming) 'Hydraulic power as a public service 1891 to 1975', *Transactions of the Institution of Engineers, Australia—Mechanical Engineering*, vol. ME22, no. 3/4.

—— and Smith, Peter R. 1983, *Environmental Systems*, Van Nostrand Reinhold, New York, 240 pp.

—— & —— 1988, *The Science and Technology of Building Materials*, Van Nostrand Reinhold, New York, 213 pp.

—— & Wilson, Forrest 1981, *Structural Systems*, Van Nostrand Reinhold, New York, 256 pp.

Cox, Philip & Freeland, John 1969, *Rude Timber Buildings in Australia*, Thames & Hudson, London, 215 pp.

—— & Lucas, Clive 1978, *Australian Colonial Architecture*, Lansdowne, Melbourne, 280 pp.

—— & Stacey, Wesley 1971, *Building Norfolk Island*, Thomas Nelson, Melbourne, 96 pp.

Cuffley, Peter 1973, *History of Oil and Kerosene Lamps in Australia*, Pioneer Design Series, Yarra Glen, Vic., 192 pp.

Daniels, Farrington 1964, *Direct Use of the Sun's Energy*, Yale University Press, New Haven, 374 pp.

de Dear, R. & Auliciems, A. 1988, 'Air conditioning in Australia—user attitudes', *Architectural Science Review*, vol. 31, pp. 19–34.

Dickinson, Tom 1893–94, 'Notes on hydraulic power supply in Sydney', *Minutes of Proceedings of the Engineers Association of NSW*, vol. 9, pp. 38–58.

Dingle, Tony & Rasmussen, Carolyn 1991, *Vital Connections: Melbourne and its Board of Works, 1891–1991*, McPhee-Gribble, Melbourne, 432 pp.

Drew, Philip 1992, *Veranda: Embracing Place*, Angus & Robertson, Sydney, 246 pp.

Dunston, David 1989, 'Electricity supply in the Sydney region', in *Sydney: From Settlement to City*, ed. D. Fraser, Engineering Heritage Committee, Sydney Division, Institution of Engineers, Australia, and Engineers Australia Pty Ltd., Sydney, pp. 273–84.

Edwards, Cecil 1969, *Brown Power*, State Electricity Commission of Victoria, Melbourne, 313 pp.

Elliott, Cecil D. 1992, *Technics and Architecture: The Development of Materials and Systems in Buildings*, MIT Press, Cambridge, Mass., 467 pp.

Fanger, P. O. 1972 (1970), *Thermal Comfort*, McGraw-Hill, New York (first published by Danish Technical Press, Copenhagen), 244 pp.

Filarete 1966 (1464), *Filarete's Treatise on Architecture*, 2 vols, ed. J. R. Spencer, Yale University Press, New Haven, 339 pp. + facsimile of MS.

Fitzpatrick, F. W. et al. 1912, *Cyclopedia of Fire Protection, Vol. II*, American School of Correspondence, Chicago, 460 pp.

Fletcher, Sir Banister 1987, *A History of Architecture*, 19th edn, Butterworths, London, 583 pp.

Fowell, Newton 1988, *The Sirius Letters: The Complete Series of Letters of Newton Fowell aboard Sirius*, ed. Nance Irvine, Fairfax Library, Sydney, 176 pp.

Fraser, Don 1985, 'Early reinforced concrete in New South Wales (1895–1915)', *Multi-disciplinary Engineering Transactions, Institution of Engineers, Australia*, vol. GE9, pp. 81–92.

—— (ed.) 1989, *Sydney: From Settlement to City*, Engineering Heritage Committee, Sydney Division, Institution of Engineers, Australia, and Engineers Australia Pty Ltd, Sydney, 328 pp.

Freeland, J. M. 1968, *Architecture in Australia: A History*, F. W. Cheshire, Melbourne, 328 pp.

Garran, Andrew 1886, *Picturesque Atlas of Australasia*, Picturesque Atlas Publishing Co., Sydney, 530 pp.

Garside, Janet & White, Deborah 1963, 'Arcades in Melbourne', *Architecture in Australia*, vol. 52, no. 2, June, pp. 80–4.

Gemmel, Warwick 1986, *And So We Graft from Six to Six: The Brickmakers of New South Wales*, Angus & Robertson, Sydney, 90 pp.

Gilbert, Christopher & Wells-Cole, Anthony 1985, *The Fashionable Fireplace*, Leeds City Art Gallery, Leeds, 82 pp.

Graf, Otto 1951, *Die Versuchsberichte des deutschen Ausschusses für Stahlbeton—Inhalte der Hefte 1 bis 102* (The Research Reports of the German Committee for Reinforced Concrete—Contents of Reports 1–102), W. Ernst & Sohn, Berlin, 42 pp.

Gyfford, E. 1972 (1806), *Designs for Elegant Cottages and Small Villas, Calculated for the Comfort and Convenience of Persons of Moderate and Ample Fortune*, facsimile reprint, Gregg International Publications, Westmead (UK) (first published by J. Taylor, London), 20 page + 26 plates.

Haddon, Robert 1908, *Australian Architecture*, George Robertson, Melbourne, 485 pp.

Haegermann, Gustav, Huber, Günter & Möll, Hans 1964, *Vom Caementum zum Spannbeton* (From Roman Concrete to Prestressed Concrete), Bauverlag, Wiesbaden, 72 + 197 + 57 pp.

Hassall, David 1969, *Reflective Insulation and the Control of Thermal Environments*, St Regis–ACI, Sydney, 34 pp.

Heath, Tom (Professor Emeritus of Architecture, Queensland University of Technology) 1996, personal communication, April.

—— & Moore, Tony 1963, 'Sydney's arcades—an historical note', *Architecture in Australia*, vol. 52, no. 2, June, pp. 85–90.

Herbert, Gilbert 1978, *Pioneers of Prefabrication: The British Contribution in the Nineteenth Century*, Johns Hopkins University Press, Baltimore, MD, 228 pp.

Herman, Morton 1970, *The Early Australian Architects and their Work*, 2nd rev. edn, Angus & Robertson, Sydney, 248 pp.

Hill, Leonard 1914, *Report on Ventilation and the Effect of Open Air and Wind on Respiratory Metabolism*, Reports of the Local Governments Board of Public Health, New Series, no. 100, London.

Houghten, F. C. & Yaglou, C. P. 1923, 'Determination of the comfort zone', *Transactions of the American Society of Heating and Ventilating Engineers*, vol. 29, p. 361.

*HRA* 1914, *Historical Records of Australia*, Series 1, *Governors' Despatches to and from England*, vol. I, 1788–1796, Parliamentary Library, Commonwealth of Australia, Melbourne, 822 pp.

*HRNSW* 1892, *Historical Records of New South Wales*, Government Printer, Sydney, 7 vols.

*HUD* 1975, *In the Bank or Up the Chimney?*, Department of Housing and Urban Development, Washington, DC, 70 pp.

Hughes, Robert 1987, *The Fatal Shore*, Collins Harvill, London, 688 pp.

ICE 1913, *Preliminary Report of the Committee on Reinforced Concrete*, 1910, and *Second Report*, 1913, Institution of Civil Engineers, London, 262 + 187 pp.

IEAust 1993, *Sydney's Engineering Heritage: Walks in the City*, Institution of Engineers, Australia, Sydney, 40 pp.

—— 1994, *The Historic Engineering Plaques of Australia*, Institution of Engineers, Australia, Sydney, 38 pp.

Ingels, Margaret 1952, *William Haviland Carrier: Father of Air Conditioning*, Country Life Press, Garden City, NY, 176 pp.

IPENZ 1994, *The First Australasian and Seventh Australian Conference on Engineering Heritage, Christchurch 27–30 November, 1994*, Institution of Professional Engineers, New Zealand, Wellington, 209 pp.

Isaac, P. G. 1958, *Roman Public Works Engineering*, Bulletin no. 13, Department of Civil Engineering, King's College, Newcastle, UK, 16 pp.

Jack, R. Ian & Cremin, Aedeen 1994, *Australia's Age of Iron: History and Archaeology*, Oxford University Press, Melbourne, 175 pp.

Jenks, Charles 1997, *The Language of Post-modern Architecture*, Academy Editions, London, 136 pp.

Joint House Department 1991, *What Makes Australia's Parliament Tick: An Illustrated Look at the Structure, Engineering Systems and Services that Make Australia's Parliament House Tick*, Australian Government Publishing Service, Canberra, 40 pp.

Jones, Bernard E. (ed.) 1913, *Cassel's Reinforced Concrete*, Waverley Book Company, London, 432 pp.

Jones, Shar 1984, *Let There Be Light*, Historic Houses Trust, Sydney, 16 pp.

Kempton, W. & L. Lutzenhiser (eds) 1992, 'Air conditioning: the interplay of technology, culture and comfort.' *Energy and Building*, special issue, vol. 18, pp. 1–2.

Kennedy, Brian & Kennedy, Barbara 1993, *Sydney Tunnels*, Kangaroo Press, Sydney, 96 pp.

Kent, H. C. 1924, 'Reminiscences of building methods in the seventies under John Young', *Architecture* (Sydney), Nov., pp. 5–13.

Laing, Alastair 1982, *Lighting: The Arts and Living*, H.M. Stationery Office, London, 72 pp.

Latta, David 1986, *Lost Glories: A Memorial to Forgotten Australian Buildings*, Angus & Robertson, Sydney, 176 pp.

Le Corbusier 1947 (1937), *When the Cathedrals Were White*, McGraw-Hill, New York, 217 pp. (originally published as *Quand les Cathédrales etaient Blondes*).

Leeper, G.W. (ed.) 1955, *Introducing Victoria*, Australian & New Zealand Association for the Advancement of Science, Melbourne, 306 pp.

Lewis, Miles 1985, 'Monier and anti-Monier: early reinforced concrete in Australia', *Second National Conference on Engineering Heritage, Melbourne, 20–22 May, 1985*, Institution of Engineers, Australia, Publication 85/3, Canberra, pp. 85–90.

—— (ed.) 1988, *200 Years of Concrete in Australia*, Concrete Institute of Australia, Sydney, 137 pp.

McGrath, Raymond & Frost, A. C. 1961, *Glass in Architecture and Decoration*, Architectural Press, London, 712 pp.

Mackay, Richard 1995, 'Picking through the pieces', *National Trust Quarterly*, February, pp. 13–14.

McLennan, I. 1974, *History of the Australian Steel Industry*, University of Queensland Press, Brisbane, 65 pp.

McMillan, R. E. 1944, *The Influence of the Lessons of the War on the Future of Town Planning, Building Design and Other Matters*, Department of Home Security, Canberra, 101 pp.

McNeil, Ian 1972, *Hydraulic Power*, Longmans, London, 197 pp.

Marquis-Kyle, R. & Walker, Meredith 1992, *The Illustrated Burra Charter*, Australia ICOMOS, Sydney, 84 pp.

Marsh, Charles F. 1904, *Reinforced Concrete*, Archibald Constable, London, 545 pp.

Martin, Ged (ed.) 1978, *The Founding of Australia*, Hale & Iremonger, Sydney, 313 pp.

Maucer, Samuel 1934, *The Rise and Progress of the Metropolitan Fire Brigade*, The Brigade, Melbourne, 42 pp.

Mitchell, Gerald Wilkinson 1922, 'Genesis and development of concrete', *Transactions of the Institution of Engineers, Australia*, Sydney, pp. 172–174.

Morgan, E. J. R. & Gilbert, S. H. 1969, *Early Adelaide Architecture 1836–1886*, Oxford University Press, Melbourne, 169 pp.

National Trust 1969, *Historic Homesteads of Australia*, 2 vols, published for the Australian Council of National Trusts by Cassell Australia, Melbourne, 280 + 320 pp.

—— 1987, *Victoria's First Government House, 1834–1854*, National Trust of Australia (Victoria), Melbourne, 16 pp.

—— 1991, *Walking Melbourne*, National Trust of Australia (Victoria), Melbourne, 68 pp.

Natshen Yusuf (Director of the Department of Antiquities, El Aqsa Mosque, Jerusalem) 1980, personal communication, July.

NEAC 1975, *Proposals for an Australian Conservation of Energy Policy*, National Energy Advisory Committee, Commonwealth Government Printer, Canberra, 34 pp.

NSW Legislative Assembly 1898, Report of the Royal Commission approved on 21st May 1896, To enquire … into any improper conduct in connection with the contracts, past and present, of Carter Gummow and Co., John Carter, D. C. Snodgrass, George Maddison, Peter Ewing, F. M. Gummow, George Forrest, and James Gillan with the Department of Public Works … Together with the Minutes of Proceedings and Evidence, Appendices, and Plates, *Votes and Proceedings of the Legislative Assembly of New South Wales during the Session of 1897*, Government Printer, Sydney, vol. 6, LXXXI + 173 + 785 pp. + 84 appendices.

O'Brien, Geraldine 1993, 'Debate rages over house that may be Australia's oldest', *Sydney Morning Herald*, 31 May, p. 10.

O'Dea, W. T. 1948, *Darkness into Daylight: An Account of Past, Present, and Future Man-made Illumination*, H.M. Stationery Office, London, 34 pp.

—— 1958, *A Short History of Lighting*, H.M. Stationery Office, London, 40 pp.

O'Farrell, Patrick (ed.) 1971, *St. Mary's Cathedral 1821–1971*, Devonshire Press for St Mary's Cathedral, Sydney, 251 pp.

Ogg, Alan 1987, *Architecture in Steel: The Australian Context*, Royal Australian Institute of Architects, Canberra, 255 pp.

Oldham, Ray & Oldham, John 1961, *Western Heritage: A Study of the Colonial Architecture of Perth, Western Australia*, Paterson Brokensha, Perth, 102 pp.

O'Neill, Judith 1974, 'The environment of tall buildings', in R. M. Aynsley & H. J. Cowan, *Proceedings of the Australian and New Zealand Conference on the Planning and Design of Tall Buildings, 14–17 August, 1973*, University of Sydney, pp. 288–314.

Oxford 1967, *The Shorter Oxford Dictionary*, 3rd rev. edn, ed. C. T. Onions, Oxford University Press, Oxford, 2515 pp.

Parsons, Nicole (Otis Elevator Company, Minto, NSW) 1996, personal communication, 19 March.

Pearson, Sue 1988, *Building Materials Analysis: First Government House Site, Sydney*, Heritage Research Services, ANUTECH Pty Ltd, Canberra, 63 pp.

Pegrum, Roger 1983, *The Bush Capital*, Hale & Iremonger, Sydney, 192 pp.

Pevsner, Nikolaus & Richards, J. M. 1973, *The Anti-Rationalists*, Architectural Press, London, 210 pp.

Prager, Frank D. & Scaglia Giustina 1970, *Brunelleschi: Studies of his Technology and Inventions*, MIT Press, Cambridge, Mass., 152 pp.

Proudfoot, Helen 1971, *Old Government House: The Building and its Landscape*, State Planning Authority of New South Wales in assoc. with Angus & Robertson, Sydney, 91 pp.

——, Bickford, Anne, Egloff, Brian, & Stocks, Robyn 1991, *Australia's First Government House*, Allen & Unwin, in conjunction with the Department of Planning, Sydney, 196 pp.

RAIA 1985, *Building Conservation in Australia*, eds Peter Freeman, Eric Martin & John Dean, Royal Australian Institute of Architects, Education Division, Canberra, 108 pp.

RIBA 1911, *Second Report of the Joint Committee on Reinforced Concrete*, Royal Institute of British Architects, London, 48 pp.

Robertson, Graeme 1967, *Ornamental Cast Iron in Melbourne*, Georgian House, Melbourne, 229 pp.

—— & Robertson, Joan 1975, *Parkville*, Georgian House, Melbourne, 184 pp.

—— & —— 1977, *Cast Iron Decoration*, Thames & Hudson, Melbourne, 336 pp.

Roxburgh, Rachel 1974, *Early Colonial Houses in New South Wales*, Ure Smith, Sydney, 603 pp.

Ruskin, John 1960, *The Stones of Venice*, ed. & abr. J. G. Links, Collins, London, 254 pp.

Saunders, David 1959, 'The reinforced concrete dome of the Melbourne Public Library, 1911', *Architectural Science Review*, vol. 2, pp. 39–46.

—— (ed.) 1966, *Historic Buildings of Victoria*, National Trust of Australia (Victoria) and Jacaranda Press, Melbourne, 278 pp.

Schlüter, F. 1942, 'Aus den Anfängen der Beton-Eisen-Bauweise' (The first attempts to build reinforced concrete), *Beton und Eisen*, no. 1/2. Quoted by Haegermann *et al.*, 1964, p. 68+.

Selfe, N. 1893–94, 'Rise and progress in lift construction in New South Wales', *Minutes of Proceedings of the Engineers Association of NSW*, vol. 9, 11 October.

Serle, Geoffrey 1982, *John Monash: A Biography*, Melbourne University Press, Melbourne, 600 pp.

Sharp, W. Watson 1969 (1946), *Australian Methods of Building Construction*, rev. edn, Angus & Robertson, Sydney, 477 pp.

Skempton, A. W. 1962–63, 'Portland cements, 1843–1847', *Transactions of the Newcomen Society*, vol. 35, pp. 117–52.

Smeaton, John 1813, *The Narrative of the Building of the Eddystone Lighthouse with Stone*, Longman, Hurst, Rees, Orme & Brown, London, 198 pp.

Spry, A. H. 1983, *Australian Building Sandstones*, Australian Mineral Development Laboratories, CSIRO, Frewville, SA, 100 pp.

—— 1986, *Australian Building Stones: Marble*, CSIRO Division of Building Research, Melbourne, 116 pp.

Stewart, Randall 1965, 'Parthenon', *Encyclopedia Britannica*, vol. 17, Encyclopaedia Inc., Chicago, p. 341.

Sumner, Ray 1985, 'The Queensland Style' in *The History of the Australian House*, ed. Robert Irving, Oxford University Press, Melbourne, pp. 291–313.

*SMH* 1898, 'Description of the Queen Victoria Market Building', *Sydney Morning Herald*, 21 July.

Tench, Captain Watkin 1961 (1793), *Sydney's First Four Years, being a reprint of* A Narrative of the Expedition to Botany Bay, *published in London in 1789, and of* A Complete Account of the Settlement of Port Jackson, *published in London in 1793*, ed. L. F. Fitzhardinge, Angus & Robertson, Sydney, 364 pp.

Thorne, Ross (Associate Professor of Architecture, University of Sydney) 1995, personal communication, May.

Thorpe, N. J. 1956, 'Centenary of Sydney's Sewage', *Sydney Water Board Journal*, vol. 6, no. 2, July, pp. 45–52.

Turner, Brian 1985, *Australia's Iron Lace*, Allen & Unwin, Sydney, 191 pp.

Villard de Honnecourt 1966, *The Sketchbook of Villard de Honnecourt*, ed. T. Bowie, Indiana University Press, Bloomington, Ind., 144 pp. (facsimile and translation of a manuscript, probably compiled between 1225 and 1250, now in the Bibliothèque Nationale, Paris).

Vitruvius Pollo, Marcus 1960, *The Ten Books of Architecture*, trans. M. Morgan, Dover, New York, 331 pp.

Wade, John 1979, *Sydney International Exhibition 1879*, Museum of Applied Arts and Sciences, Sydney, 56 pp.

Ware, Dora 1967, *A Short Dictionary of British Architects*, Allen & Unwin, London, 312 pp.

Warren, William Henry 1902, 'Investigation in regard to concrete with steel rods', *Journal of the Royal Society of New South Wales*, vol. 36, pp. 290–313.

—— 1904, 'Further experiments on the strength and elasticity of reinforced concrete', *Journal of the Royal Society of New South Wales*, vol. 38, pp. 140–169.

—— 1905, 'Reinforced concrete', *Journal of the Royal Society of New South Wales*, vol. 39, pp. 49–64.

—— 1921, *Engineering Construction: Part II, In Masonry and Concrete*, Longmans, Green & Co., London, 498 pp.

Watson, A. Frederick 1911, *A History of the Sydney Hospital from 1811 to 1911*, W. A. Gullick, Government Printer, Sydney, 212 pp.

Wells, S. 1975, 'Edison's electric light', *IES Lighting Review* (Sydney), vol. 37, Aug., pp. 97–9.

Wingler, H. M. 1969, *The Bauhaus*, MIT Press, Cambridge, Mass., 653 pp.

Winslow, W. M. 1963, *A Libation to the Gods*, Hodder & Stoughton, London, 191 pp.

Woolrich, W. R. 1969, 'The history of refrigeration: 220 years of mechanical and chemical cold, 1748–1968', *ASHRAE Journal*, vol. 11, July, p. 34.

Xenophon 1968, *Memorabilia and Oeconomicus*, trans. E. C. Merchant, Harvard University Press, Cambridge, Mass., 525 pp.

*Yearbook of the Commonwealth of Australia*, 1960, no. 46, Commonwealth Bureau of Census and Statistics, Canberra, 1269 pp.

# INDEX